THE COMING
OF THE
DRYAS

Confessions of a Mad Scientist:
Global Warming & How I Did It
With Nicholas Thornbro

The Coming of the Dryas (first edition, 2012):
Amidst the Warming a Cold Weather Clock
Ticks Away in the Northern Seas

The Would-be Adventures of Pub and Indy:
A Discourse on the Public Mind

An Uncertain Justice:
Examination of the Eyewitness and Photographic
Evidence in the Assassination of John F. Kennedy

In the Footsteps of Yeshua:
An Ancient Guide to Living

Guardians of the Grail:
The Promise of an Ancient Wisdom
'A Drink from the Bubbling Spring'

A Midsummer's Dream:
For Thought and Inspiration

Poems for Kids:
For Childhood and Youth

Where the Wild Blueberries Grow:
Reflections of the Heart

THE COMING

OF THE

DRYAS

A COLD WEATHER CLOCK TICKS
AWAY IN THE NORTHERN SEAS

Second (Updated) Edition

WILLIAM THORNBRO
&
NICHOLAS THORNBRO

ISBN: 979-8-89031-865-7 (sc)
ISBN: 979-8-89031-866-4 (hc)
ISBN: 979-8-89031-867-1 (e)

One Galleria Blvd., Suite 1900, Metairie, LA 70001
(504) 702-6708

To Christopher
(the weatherman)

With one great change upon us,
Another greater is at hand.
The Atlantic Current ceasing,
Brings cold to the northern lands.

It's in the geologic record-
This has happened in the past.
Before the Earth's the same again,
Many thousand years might pass.

'Tis a call to all hearts, I trumpet!
Let the quest for the future begin.
Sun wind and atom will sustain us;
There's a challenge to meet,
A world to win!

A Call to All Hearts,
Where the Wild Blueberries Grow, 2021

CONTENTS

Part One
Recent Events- Profound Change in The Northern Seas

Part Two
The Past as Prologue

Part Three

Unprecedented Change

Part Four

Threshold to a Brave New World

ILLUSTRATIONS

PREFACE TO THIS SECOND EDITION

The inspiration for this project, aside from a long-standing interest in climate, has occurred with the revelation in recent years and decades of a weakening in the Atlantic's deep southbound current and a subsequent finding of two distinct climate shifts in the North Atlantic and Subpolar Seas. These events have a significant impact on the climate system in the Arctic as well as globally, especially in the North Atlantic region. Analysis of cyclical changes in certain orbital conditions of the Earth and the far-reaching climate changes they have produced in the past, and considering the rapidity with which some of them are found to occur- all of this reveals an increasing relevancy of climate study to contemporary societies. An understanding of climate and how it is likely to change are, in fact, vital to the survival and the success of humanity, and it is ultimately toward this end that this work is dedicated.

This study's content, although publicly available, is not well known outside scientific circles; its assimilation includes, as its primary basis, a wealth of published research by climatologists, geophysicists, and others in related fields, spanning, in particular, the period of the last several decades. Sources are cited in the text by the author's name, title of a research paper, or the name of a website, along with the date of publication- with the associated list of references included at the end; this is in keeping with the customary notation format used in the field of climatology, and other associated scientific literature. Page numbers included in the list of references are to indicate a work's location within a larger publication, such as a periodical or symposium. The reader is encouraged to spend some time with entire articles or research papers so as to get their full benefit and to see each source in its intended context.

While there are a number of points referenced that may have many sources, we have endeavored to avoid overpopulating the text with notes; there are a few citations, however, that list more than one source. A glossary (terms and definitions) is included to clarify some words that may be unfamiliar. This edition includes, in some instances, clarifications of terms within the text itself in order to improve their readability, and due to their importance and complex interactions, descriptions of some climate aspects and their processes will be found in more than one location. Additionally, there are a few new terms that we have, for purposes of discussion, taken the liberty to coin. Metric measurements are used throughout the book, in addition to their English system equivalents. A timeline is added to display the changes in the northern seas, as well as a numerical index highlighting events in recent decades in the North Atlantic Oscillation (NAO).

There are some surprising revelations with regard to the course we can expect the climate to take. The evidence, in particular from the climate record- that is, the record of the past- strongly suggests that trace greenhouse gasses- namely, carbon dioxide, methane, and so on- do not play a determining role in climate change and that we should be concerned that the current warming, rather than leading to an increasingly warmer future, may culminate in a greater change towards much cooler climate conditions. The conclusions drawn are solely those of the authors of this work, as non-scientist writers, and are generally found to be contrary to prevailing opinion, voiced by official sources and mainstream media. They never-the-less represent at least the basic views of what may be called a significant minority among climatologists and the scientific community as a whole.

INTRODUCTION

Observations of the climate in the North Atlantic and in the Subpolar Seas, with data obtained from instrumental measurements, satellite Altimetry, samples taken from glacial deposits, deep sea cores, and lake beds, and from historical records, reveal critical changes in atmospheric and oceanic circulation. These changes have become increasingly evident in the last several decades, particularly since the early 1990s. Analysis indicates a progressive weakening of the climate system, with a significant shift centering on the early winter months of 1996, followed fourteen years later by a second and more substantial shift beginning in 2010. Changes have occurred in the relative influence of Arctic and Atlantic water masses. There is a large increase in the volume of Arctic-derived water and a corresponding decrease in the strength of the North Atlantic Current. Paramount among the changes are a declining strength of ocean circulation in the Subpolar Seas, including a slowing of the Subpolar Gyre, a sharp decline in the critical deep-water convection process, and a resulting weakening at the surface, as well as in the deep southbound current. This region is found to be especially sensitive to climate change, an important precursor to global change, and a critical link in the far-reaching circulation system of the world's oceans.

In the broadest sense, these events can be understood within the context of long-term latitudinal changes in direct sunlight, or Insolation, determined by cyclical changes in the precession and inclination of the Earth's axis, its proximity to the Sun, and the shape of its orbit. On a more immediate basis, though, they are attributed to changes and increasing variability in atmospheric conditions, oceanic changes due to Arctic warming, and an accelerated Hydrological Cycle- all of which directly proceed from an increasingly intense tropical warming.

The implications for human societies are immense. Climate change is bringing about reductions in the habitable range of many land areas, presenting new challenges in the distribution of populations, energy, and even the most basic of needs. The dangers posed by the current warming are increasingly evident; there are, however, extraordinary and opposing climate signals occurring in the North Atlantic area and also globally that precede significant change in the northern regions and beyond. As displayed on the cover, we are today, figuratively speaking, counting down the final minutes of the warm climate conditions we have known. A cold weather clock ticks away in the northern seas that portends the coming of a new age of snow and ice.

Wm. Thornbro, December 2023

PART ONE

RECENT EVENTS- PROFOUND CHANGE IN THE NORTHERN SEAS

1

AN OVERVIEW OF HOLOCENE CLIMATE CHANGE

With a look toward providing a broad perspective- a background of sorts - within which some of the more challenging aspects of climate change can be better comprehended, the following pages will serve as a bit of a preview of some essential highlights since the end of the last ice age.

Studies over the last several decades show a progressive weakening (meaning less vigorous) of ocean circulation in the North Atlantic and north of there in the Nordic Seas. The reason this is important is because ocean currents in the Atlantic connect to a worldwide system of surface and deep-water currents, which are vital to the distribution of heat, oxygen, and nutrients throughout the oceans. A weakening in the Atlantic link produces a reverberating effect downstream in the other oceans as well. An understanding of the importance of ocean life to us land dwellers should be enough to gain our complete attention, but aside from that, the geologic record- the record of the climate of the past-associates such events with sweeping global change.

Following a period of particularly warm conditions, a 'climate optimum,' some fifty million years ago, the Earth's climate has steadily cooled. Such a concept may sound strange, considering that today's talk is about warming, but it's a fact. Ice age conditions have been the prevailing climate state for the last 33 million years. Moreover, for the last 1.5 million years, the Earth has alternated between long periods of those ice age conditions and relatively short intervals of warmer climate- like what we have today. Long-term cycles in the angle of the Earth's axis, the shape of its orbit, and its proximity to the Sun produce changes in the distribution of solar energy- called 'Insolation'- on the Earth's surface. These conditions are responsible for the alternating change between cold glacial and warm interglacial climates. When these climate factors come together in the Northern Hemisphere, with its great expanse of land masses, unique geographical distribution and capacity to retain heat, the Earth warms up to conditions like today.

At the beginning of this warming period, called the Holocene, and before then, some 15,000 years ago, the focus of the Sun's warmth was in the north. Today, those cycles responsible for the initial warming have moved away from those ideal positions. Northern temperatures have declined since the initial warming in the early Holocene as the focus of the Sun's energy, of Insolation, has gradually moved south- its warmth is now most directly focused on the Southern Hemisphere.

Thus, we find ourselves today, as a result of these changes, at a place where those great parameters of change are positioned to return the climate to glacial conditions. The actual transition, when it occurs, will be determined by events going on on the Earth's surface. Overall, sea surface temperatures since the very beginning of the Holocene have trended downward in the polar regions and increased at tropical latitudes. Due to changes in Insolation, the contrast is most striking in the Northern Hemisphere. As a result, there is an acceleration of a natural tropic-to-pole movement of air and everything else that goes along with it.

Consequently, due to increasing tropical warming, evaporation is on the rise, too. All elements, therefore subject to evaporation, are increasingly accumulating in the atmosphere, with much of it transferred toward the poles. The most important in this process is water. Its presence in the air in its evaporated water vapor form is known to have grown by at least 10% since 1900. It is called the acceleration of the Hydrological Cycle, which essentially involves the cycling of moisture- of water- through the environment. It is global in scope and of particular importance in this revving up of climate conditions.

As you might imagine, this creates some problems north and south of the Equator. The direction of flow in the oceans and atmosphere is always away from the tropics toward the poles. More water in the air means a more active environment, with more rain, storms, and flooding. It also means less water someplace else. It is not just in the tropics that evaporation is increasing; to a lesser extent, it is increasing worldwide. The land is becoming drier as more and more water is taken up in atmospheric processes, giving rise to extensive and widespread burning of forest lands-wildfires.

Because of its importance in large climate change, we need to bring special attention to what is going on in the Arctic. A good portion of the extra water, due to the acceleration of the Hydrological Cycle, ends up there in the form of added precipitation- snow and rain. And along with the air and water comes the tropical heat. Much of the Arctic warming, therefore, rather than being simply due to an overall global effect, has, in fact, a tropical origin.

The Arctic Ocean is, for the most part, surrounded by land, and rather like a bathtub when enough water is added, it will overflow. The only outlet for it is southward to the Atlantic. Because of the addition of all this extra water, there has been, since the beginning of the 1950s, a growing awareness of an increasing flow of cold and fresh Arctic water, mainly through the Nordic Seas to the Atlantic and beyond.

The study of climate and how it has changed in the past reveals this same area, the Nordic Seas and North Atlantic, to be the place where ice ages begin and end. From far to the south flows a great river of warm, salty Atlantic water; the true champion of global warmth, we call the Atlantic Current. It plays a critical role in the unfolding drama. In effect, it is the one single lifeline for this entire period of warm interglacial climate. Without it, there is no warm climate- there is, in its place, an ice age.

After some 15,000 years of warming, our present period of interglacial warmth has lasted for about as long as we can expect one to last. And there is no reason to think that the climate will not follow the same cycle of change that it has taken in the past. Large climate change occurs over hundreds and even thousands of years, but, as we will later see, this process is already underway and has been for hundreds and, in some respects, thousands of years. Realistically, no one knows for sure how long the change will take. While it may require some lengthy time to complete, recent events demonstrate that, rather like the quick movement in an earthquake, similar sudden shifts in the climate along the way are likely to produce profound difficulties.

Two such shifts, in fact, have already occurred- in the mid-nineties and beginning in 2010. Among the most visible and recent changes are the weakening of the Jet Streams. The strength of these once great west -to-east global currents of air in the mid-latitudes has declined considerably, along with their associated polar vortexes. Increased glacial melting is especially evident in the coastal areas of Greenland, Iceland, Norway, and elsewhere, and there is a general sea ice loss as part of increasing warming. These aspects of climate in the Arctic have gotten much attention. What has yet to get much attention, however, and which bears much responsibility for the deteriorating conditions that are evident there, is an increasing southward flow of Arctic water- the culprit being all that extra water introduced into the Arctic.

The history of climate change in the Atlantic and northern seas can be seen as a long and enduring struggle. It is an ages-old push-and-pull contest between the Arctic and Atlantic: cold against warm- freshwater against saline. The defining events of the Holocene warming period can be understood within the context of the relative balance of Arctic and Atlantic water in this region. A brief look at that balance will underscore its importance and provide a good idea of where we are today in the scheme of things and where things are headed.

The Earth's oceans consist of three primary layers. Some mixing between them occasionally occurs, such as during stormy winter seasons, but otherwise, all tend to remain unmixed- maintaining their chemical properties. The same is true with ocean currents- those great rivers of water that flow through all of the oceans- and water masses in general. Most important to the climate are temperature and salinity levels. Cold and fresh water from the Arctic and the warm salty tropical-derived water that comes north in the Atlantic Current come into contact with one another in this region but otherwise tend to stay to themselves- you might say.

During the last ice age, there would have been a near-total dominance of Arctic water. Then, after 100,000 years of cold, as temperatures first began to rise above glacial levels about 15,000 years ago, there was an increasing presence of Atlantic water. The Holocene began with the most rapid warming on record, some 11,500 years ago, and by 10,000 years ago, temperatures had become relatively stable near today's levels. By 9,000 years ago, the ice covering the Nordic Seas had almost completely disappeared, and Atlantic water had become hugely dominant. At the 7,000-year marker, unmixed Atlantic water had reached its greatest extent, occupying nearly the entire width of the Nordic Seas region, all the way into the Arctic. It was the time of the absolute height of warmth and humidity of this interglacial period. The whole world was warmer and greener with vegetation than today. By 5,000 years ago- the beginning of the 'Late Holocene'- things had begun to change. It marked a final transition from the early height of Holocene warming to a period of progressive decline. The climate had begun to become a little cooler and dryer. Deserts, which had largely disappeared during the early period, began to expand. The northern extent of Atlantic water scaled back to about the mid-point in the Nordic Seas. And fast on its heels, pressing ever southward was an expanding presence of Arctic water. By 3,000 years ago, Atlantic water had become restricted to a small area in the far southeastern area of the Nordic Seas. Since then, the climate has continued to cool down. Over the last several thousand years, temperatures have increasingly oscillated back and forth between warm and cold conditions, culminating some 670 years ago in a 'Little Ice Age'- 500 years of the coldest climate since the warming began.

The end of that 'Little Ice Age' and the beginning of today's warming occurred in about 1850. And while a southbound flow of Arctic water through the Atlantic and beyond has long been a natural feature of ocean circulation, today, we find ample evidence of its unabated and increasing influence, extending all the way south into the Southern Ocean around Antarctica- largely the result of the acceleration of climate processes. In the primary area of concern, the North Atlantic and Nordic Seas, since the 1980s, the precipitation rate, as a result of the changing water cycle, has increased at least thirty percent.

In order to maintain that critical link in global ocean circulation, the Atlantic Current, at some point, has to do a turn-around and head back south to connect up with other currents. There are two primary places where that turn-around occurs; the first and most important is the Greenland Sea, and the second is the Labrador Sea. The way it works is in a process called convection. Atlantic water is the most saline of all the oceans, and entering the Greenland Sea region, it has cooled; much heat has evaporated off the surface- especially when strong winds are present. As a result, it is made heavier than the surrounding water, and it sinks into the ocean. At the bottom of the Greenland Sea lies an ocean basin that is the deepest in the Northern Hemisphere, and there is, at its bottom, an additional fourth water layer that is the coldest in the Northern Hemisphere. Under certain ideal conditions, when surface winds are strong and contrasts between water from the Atlantic Current and the surrounding water are greatest, the Atlantic water sinks all the way to this bottom layer. It begins a journey back southward, forming into what is referred to today as Lower North Atlantic Deep Water (LNADW), which we will refer to here simply as the Deep Water Current- or Deep Current.

The reason for singling out the Greenland Sea in this is because most of the water comprising that Deep Current has come from there. It is the single most crucial link in the Atlantic region that eventually connects up to the global system of currents. It has long been believed that an interglacial climate- like what we have today- cannot be maintained in the absence of a strong formation of deep water from the Greenland Sea. Nowadays included as a part of what is called the Atlantic Meridional Overturning Circulation (AMOC), the strength of the Deep Current has been continuously monitored- since 2004, and before that, periodic measurements were made beginning in 1957- all of which reveal a long -term decline. The convection process occurs elsewhere in the world as well; in some places, the water moves upward instead of downward as it connects to other ocean currents. However, nowhere else is there a comparable volume of this sinking-down-welling- process.

One particularly curious aspect of the climate is that an event going on in one place is often found to be balanced by another event of opposite sign- in other words, opposite effect- someplace else. As our modern period of warming began to rev up to a higher gear in the 1980s, measurements in the Greenland Sea began to reveal a progressive weakening of the convection process, brought on by more and more mixing and replacement in the convection regions by an ever-increasing in-flow of Arctic water through the surface and deep layers. A defining event of the mid-nineties shift involves about an 80 percent decline in the Greenland Sea contribution to the Deep Current.

In the Labrador Sea, the second-most important location, situated in the far northwest Atlantic, there is little or no direct contribution to the Deep Current. Named the Upper North Atlantic Deep Water (UNADW), its position is directly above the Deep Current on its southward journey. It does not connect up to other currents, however- rising to the surface off the coast of West Africa. Nevertheless, by the mid-nineties, due to the ongoing invasive replacement of Atlantic water by that of Arctic origin, the convection process in that sea was also in rapid decline.

The Winter of '96 saw a record negative shift in the North Atlantic Oscillation (NAO)- an important climate indicator described in Chapter 2. Warm water sea-surface plankton, dominant in the region since the beginning of the Holocene, was almost totally swept away, and in its place came a new dominance of cold water species- in itself, it is an essential indicator of the direction in which the climate is headed. A check in 2004 showed the volume of the southbound Deep Water Current to be less than half its level when measurements began in the late 50s.

This brings us to 2010 and an even more substantial shift- and this time around, the change came in the ocean's surface layer in the North Atlantic. A significant 30% weakening in the North Atlantic Current that winter, which had begun the year before, was an expected and direct consequence of the decline in the Deep Water Current over the preceding decades. What followed in the ensuing years was nothing less than a redistribution of Atlantic and Arctic water in the North Atlantic region. As the Atlantic Current weakened, there began a historic, absolutely huge formation of Arctic water in the North Atlantic. By 2016, the central mass of its accumulation had settled in the far northeast Atlantic - called the 'Eastern Subpolar Gyre'- directly in the path of a weakened Atlantic Current. Its influence included nearly the entire width of the North Atlantic Basin, from Iceland in the north to about 40° N latitude farther south. In the years following its appearance, it extended down more than 1,000 meters (some 3,300 feet)- nearly to the bottom of the surface layer. For the first time during the Holocene Interglacial Period, Arctic water has become dominant in the North Atlantic. In a little over a year after 2010, the Atlantic Current recovered but then began a new, slow, and continuing decline, having so far lost about half of what it had regained. Pushed south and eastward around this great Arctic water mass, for the first time this interglacial period, the Atlantic Current finds itself relegated to a secondary position in the North Atlantic, confined to a relatively narrow pathway along the European continental margin as it continues its northward journey to the Nordic Seas and beyond. It is Arctic water, you could say, that now occupies the seat at the head of the table in the North Atlantic. Having been assigned early on the name 'Atlantic Cold Blob,' the climate record shows this kind of event, the establishment of a large fresh Arctic mass of water in this part of the Atlantic- in the Eastern Subpolar Gyre- in the past to have led to the onset of ice age conditions.

Atlantic water and its tropical warmth still finds its way far to the north and into the Arctic Ocean, but its strength is not anything like it once was. And since this new weakening at the surface of the Atlantic Current and the appearance of the Blob, we now have a much more visible aspect to its declining influence. An increasing amount of it, apparent since 2004, is found to be recirculating back south into the subtropical region- the Subtropical Gyre. This is suggestive of a slowdown in both directions - north and south. This is an accumulation that might help to account for the Blob's massive presence. It means cooler temperatures going north, as well as warmer water to the south, bringing on more extensive and stronger storm activity.

If there is any good news in all this for an embattled Atlantic Current, it might be that its long struggle to hold back against a growing Arctic menace may be nearing its end. This ongoing and increasing re-circulation may signal the beginning of its re-organization. Along with all the recent drama in the ocean's surface layer in the region, there is an upward expansion of the deep ocean from below it and a corresponding rising of a boundary area between them- called the Thermocline. While part of today's recirculating Atlantic water turns back south at the surface, rejoining the subtropical circulation, an increasing amount of it enters that Thermocline boundary area, which allows the Atlantic water to slip through between the upper layer and the deep colder ocean, taking a deeper return route back south. Remember, bodies of water tend to stick together- maintaining their own characteristic temperature and salinity levels. It may be the beginning of the next big shift in the climate- a mid-ocean re-establishment of the overturning process, largely lost as a result of the events that have been going on farther north.

Since the changes in 2010, there has been a new acknowledgment and increasing public awareness of the recent weakening of the Atlantic Current- the AMOC. News reports increasingly focus on expectations of its collapse and when that might occur, with speculations ranging from the next century to just decades from now. But the present configuration of Arctic and Atlantic water masses in the North Atlantic is anything but stable. Expectations regarding the Blob are that it might remain until mid-century, but that too might be just wishful thinking. The Atlantic Current continues to decline, and it is difficult to imagine it holding on even to its present position in the North Atlantic, let alone the Blob just going away, in such a scenario. In the now thousands of years long history of the retreat of Atlantic water from the north, there are no instances where Arctic water has ever given up much of anything in the advancing sphere of its domain. Looking at things from a somewhat broader perspective, given the fact that solar radiance- the cycle of Insolation- is at a low point in northern latitudes- it seems doubtful that enough energy is retained within the climate system in the northern seas to effect any reversal in this long-term decline.

Realistically, the climate is always in some state of change. The Little Ice Age that preceded the modern warming was undoubtedly the coldest of the Holocene- so far. The climate, particularly in the last several thousand years, has changed back and forth between cooling and warming. There is something a lot different in the mix, however, this time around. If you figure the whole region into the equation, you have to conclude that the expected collapse is already underway and has been for at least several decades now. The events described here have not occurred since the end of the last ice age. We are at the point of a game-changing event.

There will likely come a winter- perhaps in the not-too-distant future- when a new and sudden shift in the North Atlantic climate will see the disappearance of much of the Atlantic Current north of the subtropical region and a further advance into the region of Arctic water. And the event could very well begin in the Arctic. There is a vast area of frigid cold and wind-driven circulating water in the Arctic north of the Canadian Archipelago called the Beaufort Gyre. Typically, it has reversed the direction of its circulation every six years or so, releasing a lot of its water into the surrounding Arctic, but it has not done so for a long time- at least a dozen years. It might be the event that brings on another great southward flood of Arctic water into the North Atlantic.

Recall that such an increase is associated with both the 1996 and 2010 events. Large increases in the flow of Arctic water into the Atlantic in the past are linked to a cooling of the climate. One event occurred about 8,200 years ago, and another much more severe, the Younger Dryas Event, happened early on in the Holocene that lasted for a thousand years. Like in the age- old riddle of the chicken and the egg, it may not always be easy to tell for sure which comes first during a climate shift, a weakening of the Atlantic Current or an advance of Arctic water, but what we do know is that both play a critical role.

The failure of the northern drive of Atlantic water will likely occur due to a re-establishment of its overturning circulation near the southern margins of the subpolar region. We might see it as a rebuilding of the vital Atlantic link in the global system of ocean currents. A system of currents as old and enduring as the Earth's motion on its axis and the beginning of the oceans themselves, and it will continue regardless of the goings-on in the North Atlantic. What is particularly interesting concerning this pattern of recirculating Atlantic water is that it is occurring at about the location where the overturning process has re-established itself in the past, at the end of previous interglacial periods. Retreating Atlantic water has now gone about as far south as it usually does at the end of a time of interglacial warming before switching to its glacial mode of circulation- a place where it will likely remain for a very long time.

To find a time like today, where the climate was winding down from a similar period of warmth, you would have to go back a very long time- about 115,000 years. With the failure of the AMOC, temperatures farther north will decline, and a true winter climate will prevail- the end of the Holocene will have come. And we can expect that it will be a very long time- some 75,000 years- before those great climate parameters come together again to produce the next time of climate warmth.

In the Nordic Seas, there is a 'Polar Front,' a boundary line of sorts of atmospheric storm activity separating warm Atlantic water from cold water of Arctic origin- at the very center of that Atlantic/Arctic water balance described earlier. From its beginning in the east, in the very early Holocene, along the coast of what is now Norway, about 10,000 years ago, as the ice melted away, this frontal system, at the height of Holocene warming some 3,000 years later, would find itself at its farthest extent to the west, along the still to some extent ice-bound shores of Greenland. Since then, it has begun its trek back eastward. None of us knows exactly how long its return trip will take, but when it finally completes its journey, this period of warm climate will have come to an end. The Polar Front lies today diagonally, east-to-west, across the seas. Its southern extent, since the mid-nineties event, now allows Arctic water its first access point to the North Atlantic east of Iceland since very early in its Holocene journey. Having finally cleared the only land barrier to its destiny on the eastern shore, between it and an ice age world, there is now only the open sea.

Dryas is the name of a little flower that grows in the Arctic tundra and, during glacial times, extends south into vast areas in the mid-latitudes. It is also the name assigned to the long period of ice age conditions that persisted prior to the Holocene. Along with the appearance of the Cold Blob and the colder temperatures it has brought to the North Atlantic, there are beginning signs of a new cooling trend and slowing of glacial melting in the Nordic Seas area. A coming colder climate will require us to make some important changes in the way we do things. And while these new developments may not signal the beginning of a new ice age, sooner or later, a new ice age will come. It would be wise to begin making some preparations for the coming of a brave new and cold 'Dryas' world.

2

THE WINTER OF 1995/96

A climate shift in the far north

We begin our investigation in the early months of 1996, in a winter that was unusually severe, with effects that were felt far beyond the North Atlantic. Extreme weather variability prevailed in large parts of the middle latitudes in the Northern Hemisphere, and changes in atmospheric temperature and circulation were noted worldwide. January blizzard conditions blanketed large areas of the eastern United States with 50 to 120 cm (20- 47 in) of snow, followed a week later by a rapid warming and heavy rains that produced widespread flooding. Central Canada, in particular, saw record-setting cold events, with February temperatures dipping to -46° C (-50° F) and wind-chills at -68° C (-90° F). An extremely unusual warming occurred a week or so later in the US central plains, where temperatures reached 21° C (70° F), climbing to as high as 38° C (100° F) in the south-central region by late February.

Wintertime temperatures across much of North America, on average, were about 4° C (7° F) below normal, with marked contrasts in precipitation. Similar averages occurred across Europe and Asia, although conditions were much dryer, a marked contrast to the 7° to 11° C (13°- 20° F) above-average temperatures and high humidity that prevailed the year before (Special Climate Summary, 1996).

The most important climate indicator for this part of the world is the North Atlantic Oscillation (NAO). A primary component of a larger system- the Arctic Oscillation (AO)- the NAO reflects the weather and climate conditions generally for all of the North Atlantic, as well as the Subpolar Seas, including the surrounding coastal land areas. The warm moist air produced in the climate processes in this region has a particular moderating effect on the weather across Europe and into western Asia that otherwise would experience the cold climate that prevails hundreds of miles farther north, and the NAO sometimes exerts a strong influence across broad areas of North America, and as far away as Siberia.

This climate indicator is based on the relative intensity of two atmospheric pressure systems: one centered in the northwest, near Greenland, and the other in the southeastern Atlantic. When it is in its (warmer) positive phase, the NAO is associated with stronger (meaning windier) atmospheric conditions, with a center of low pressure, typically located south and west of Iceland in winter, and a high-pressure center near the Azores. It also exhibits a (colder) negative phase where these two pressure systems become positioned farther to the south and west of their 'high' locations. When a negative phase is dominant, weather conditions become more stormy- more severe. They are also considered weaker, with strong westerly winds necessary for the transfer of heat from the ocean surface to the atmosphere shifted about 10° latitude southward and away from the far north where they are most needed to facilitate vigorous ocean circulation (Osborn, 2000).

Although the NAO fluctuates on a daily and monthly basis and exhibits at least a decadal cyclical trend, what we have seen in the last 30 to 40 years, particularly the period from the early 70s to the mid-90s, more or less in step with the generally warming climate, is an amplified NAO positive phase- the most intense on record, and perhaps the most intense in at least several hundred years. Its peak occurred in the winter of 1994/95, as the flow of relatively fresh Arctic water that had been increasing in the region for several decades neared a peak. In the months leading up to what turned out to be a significant shift in late 1995, the NAO trended strongly negative, beginning an extraordinary winter season for the North Atlantic climate.

A winter of change in the grand central station of ocean currents

The decade of the 1990s may be remembered as a time of significant climate change, with the actual turning point occurring mainly in the winter months of 1996. This was an episodic winter that saw the enhancement of essential changes that had begun several years earlier and the initiation of new ones, all of which we are still observing today. For purposes of discussion in this analysis, a reference to the North Atlantic region will include the area of the Northern Atlantic as well as the Nordic Seas (basically the area encompassing all of the Subpolar Seas), and two places can be said to be most important in maintaining today's climate conditions in this region, and beyond. The first is the Greenland Sea, where a convection process, found to be a critical link in ocean circulation, began to weaken in the early 70s, nearly ending at depth by the mid-90s. As the strength of deep ocean circulation diminished in the Greenland Sea, it began to weaken in the Labrador Sea as well- the second of these two critical areas. By the winter of '96, the convection process in this sea was also in sharp decline. A large pulse of relatively fresh water expanded southward from the Arctic that winter after a decades-long increase. Significant sea-level changes occurred in the North Atlantic, and the Atlantic Storm Tract shifted eastward as the entire climate system in the North Atlantic began a process of weakening (meaning less vigorous circulation) and contraction that would unfold over the next two years.

Meanwhile, a significant reduction became apparent in a great southbound Deep Water Current, which has global implications, for it is this current that is responsible for the completion of an Atlantic link in a worldwide ocean circulation system. From the previous historic high, the climate turned to the most sharply negative (colder, weaker) phase on record. Subsequent winters show a recovery in the NAO index, but the region's climate has not returned to its previous strength (meaning warmer, more vigorous circulation). An analysis of annual and monthly averages in the NAO index shows that since late 1995, a negative phase has been dominant; this is to say, the climate is now, most of the time, in a negative phase- reflecting a trend not seen since continuous records on this began in about 1825 (Hurrell, 1995; Jones, 1997; North et al., 2011).

The closely related Arctic Oscillation (AO), which reflects conditions for all of the Northern Hemisphere, has also trended more negative since 1996, as have temperatures in the Atlantic Current. The NAO that winter was persistently negative, as was a few years later, the winter of 2000/01. A new record low was set in the winter of 2009/10, leading in some areas to much colder weather and heavier snow cover, and a significant increase of Arctic water in the North Atlantic- which we will spend more time with later on.

This area has been called the Grand Central Station of ocean water masses and currents. Warm tropical water moves northward through the Gulf Stream in the western Atlantic, turning northeastward when it reaches the latitude of Newfoundland. From that point, it is called the North Atlantic Current, and it flows on into the Iceland Basin region in the northeast Atlantic. From the Iceland Basin, a significant part of it moves on north into the Norwegian and Greenland Seas- the Nordic Seas- and then on to the Arctic Ocean from there. Another part of the current, however, turns westward from the basin region toward the Irminger and Labrador Seas- the western extent of a region of the far northern ocean called the Subpolar North Atlantic, which extends from Europe in the east to North America in the west.

Including the Iceland Sea, five subpolar seas are important for the process of overturning and ventilation of the world's oceans. In each of the seas, broad surface water areas exhibit a slow cyclonic (counter-clockwise) circulation. When westerly winds are present and strong enough to take up sufficient moisture and heat from the ocean's surface, a convection process occurs where the in-flowing Atlantic water, now colder and more saline and therefore more dense, sinks into the ocean's water column- sometimes to great depths. As mentioned in Chapter 1, the water flows back south, completing a process referred to in scientific research as the 'Atlantic Meridional Overturning' or 'Thermohaline Circulation' (or AMOC).

The convection process may involve an area that is down-welling, where water from a surface current sinks to form an underwater current at intermediate and lower depths, such as occurs in the North Atlantic region, or it may involve an up-welling or rising of an underwater current to the surface. This process forms the connecting link between the deep and surface currents that flow through the oceans, with the North Atlantic region accounting for about 25% of the total global volume. It occurs in only a few locations throughout the world, however, and this region is unique in that there is no other location where a down-welling convection process occurs on as grand a scale. It is in the deep basins of the Greenland and Labrador Seas that the convection process reaches its greatest extent- or greatest depth.

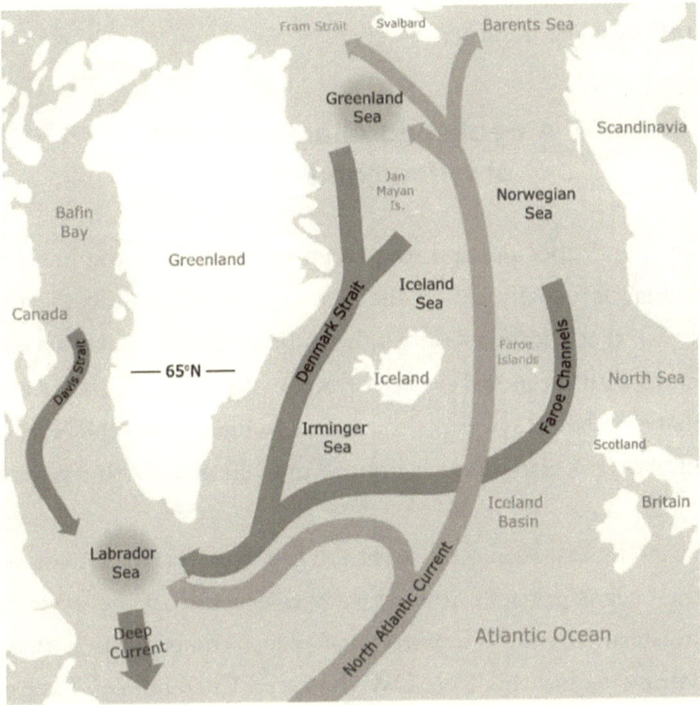

Ocean circulation in the northern seas: The Greenland and Labrador Seas being the most critical for the convection process. Note that the Iceland Basin is located in the northeast Atlantic- not to be confused with the Iceland Sea, which lies directly north of Iceland.

From the convection regions in the Nordic Seas, this Atlantic water begins its return trip southward, making its way through a system of deep water basins and channels, and enters the North Atlantic, primarily through the Faroe Channels east of Iceland. Increasing in volume as it goes, it continues to follow the contours of subterranean ridges and continental margins as it makes its way westward through the subpolar region. Eventually reaching the area of the Labrador Sea, this Deep Water Current continues, turning southward directly below a similar current of Atlantic water that comes from the Labrador Sea convection region. Each of the two maintains its own characteristic levels of temperature and salinity. From this point, they are collectively referred to as the Deep Western Boundary Current (DWBC). Both flow back toward the tropics, but the upper current from the Labrador Sea ends, surfacing in the tropical ocean off the coast of Africa.

A weakened Atlantic Current- at surface and at depth

It is only the deeper portion, the Deep Current from the Nordic Seas, which today is usually called the Lower North Atlantic Deep Water (LNADW), that continues on south to the Antarctic and completes the connection with the global system of ocean currents. It is a critical link in the global circulation system of the oceans, which makes the Nordic Seas, its point of origin, also uniquely important to climate- both regionally and worldwide.

Due to the eastward rotation of the Earth on its axis, major currents in all the oceans gravitate toward the western side of the ocean basins, so the western Atlantic is a pretty busy place. To keep things simple, we will continue to call this LNADW the Deep Current. You have got the northward flowing 'Atlantic Current' at the surface- what is called at that point, the Gulf Stream, and below that, the southward flowing current at intermediate depths that originates in the Labrador Seas region, and then the Deep Water Current below that- also flowing south.

In the Nordic Seas, the most important area for convection and the formation of the Deep Water Current has been the Greenland Sea, where the process weakened beginning in 1972, nearly ending at depth by 1994- an important climate signal. Climate data prior to '94 shows a relatively steady and growing strength of convection in the Labrador Sea, second only to the Greenland Sea in its importance in this process. The Labrador Sea, in fact, was at its height- ocean circulation and convection were the strongest ever observed- from 1990 to '94, in sync with the strong NAO positive phase. By 1994, though, Labrador Sea circulation had begun to slow; atmospheric conditions, by then, were also in a downward trend. Considerable oceanic changes can be observed beginning in the winter of '96, when the circulation system in the subpolar region, and specifically the Labrador Sea, began a steep decline. In October of that year, instrumental measurements taken in the western Labrador Sea, at the beginning of the Deep Western Boundary Current, confirmed satellite data; as the climate turned negative, convection had weakened as much as 25%, and the downward trend had occurred all the way down- through the entire water column (Häkkinen & Rhines, 2004).

Its journey takes the Deep Water Current nearly to Antarctica before it connects up with the rest of the global system. It will be shown in succeeding chapters to be an absolutely vital link in ocean circulation, and critical to maintaining today's relatively warm interglacial climate. In the Subtropical Atlantic, scientists have monitored the strength of the deep, intermediate, and surface currents since 1957 in a system of measurements spanning the width of the subtropical ocean basin, from the Bahamas to North Africa, at about 25° North Latitude. The data shows its strength to have been in decline for decades, and a 1998 measurement revealed dramatic changes. Its strength had declined to 50% of its 1957 value, with a significant amount of that change- about 44%- showing up in the '98 data, reflecting the time that includes the NAO change. The bottom 400 m (1.300 ft) of this current in 1992, was gone in '98, and still absent in subsequent measurements (Bryden, 2005).

Besides the changes at depth, important events were also happening at the surface, in the speed and direction of the northward-flowing Atlantic water. Part of the Gulf Stream- downstream of about 60° west- was found to have accelerated, while the northern branches of the Atlantic Current slowed; this is particularly evident in the area of the Iceland Basin and north of there in the Norwegian Sea (Esselborn, 1999). A slowing of the current was apparent in 1993, and it severely weakened beginning in January of '96. In the Norwegian Sea, it slowed by about 12% by 2005, and its core temperature increased by almost 1° C (nearly 2° F), a significant increase in sea-surface temperature (SST). Following the NAO change, eastern aspects of the surface current in the northeast Atlantic were weakened to the extent that they were, for a time, shifted eastward- folded-into the North Sea (Greiner, 2006). This current transports the equivalent of some 80 times the flow of the Amazon River into the Subpolar Atlantic, the Nordic Seas, and onto the Arctic Ocean, and its weakening causes a corresponding weakening downstream in each of these other areas as well.

Sea-to-air heat transfer
named an accessory to the crime

A vital aspect of the changing atmospheric conditions involves the rate of heat transfer- also called heat flux- or evaporation, from the ocean surface to the atmosphere. When the NAO changes to a negative phase, strong winds essential for the evaporation of moisture from the ocean to the atmosphere are shifted about ten degrees latitude farther south in the Atlantic. A resulting weakening of the heat transfer rate was evident in the subpolar region beginning in 1992. With the abrupt change in atmospheric circulation in '96, including an eastward shift in the Atlantic Storm Track and the accompanying southward shift of the westerly winds, there was a drastic reduction in the amount of heat and moisture being added to the atmosphere. As a result the surface waters in the far north do not reach the lower temperatures and higher salinity levels, and the resulting densities that are optimal to facilitate a robust convection process. So, along with the convection process, the heat transfer rate was also in sharp decline.

An increasing presence of Arctic water is the main culprit, but this weakening sea-to-air exchange of heat and moisture has to have also played a role- as an accessory to the crime. There are many aspects to consider in any attempt to understand the workings of the climate system and how it changes. A definite cause-and-effect relationship is evident among these important climate processes. Its every single aspect, in some way, has a direct effect on every other single aspect. And the implications extend far beyond the North Atlantic region. The high-latitude European climate, being basically downwind from the North Atlantic, is especially dependent upon the warm, moist air produced in this climate process (Esselborn, 1999).

Atlantic water-
all dressed up and no place to go

The mid-90s decline of the convection process and the accompanying new strongly negative phase in atmospheric conditions, set off a cascade of events going south. With less warm Atlantic water being drawn into the convection regions, it began to accumulate- especially in the subpolar region. General warming has been evident in the surface layer since 1993. Arriving Atlantic water, extra warm and salty from its tropical experience- all dressed up for the dance, you might say- found itself left out and with no place else to go. A resulting accumulation would occupy the entire far North Atlantic for some years to come. As we shall later see, a similar event would occur with the appearance of the 'Cold Blob' after 2010.

There are several things that can be seen to have contributed to this new mid-90s trend. Today's conventional wisdom, I am sure, will simply attribute it to 'Global Warming,' but there is plenty of evidence to indicate that there is at least a little more to it than that. In the broadest sense, today's warming, most intense in tropical latitudes, as discussed in Chapter One, plays an important role. Rising temperatures increase the evaporation rate in the tropical oceans, and since salt does not evaporate, the surface is left more saline. So warmer and more saline water is delivered from the tropical Atlantic to the far north through the Gulf Stream and the North Atlantic Current. Consequently, just as a result of this one thing alone, all areas subject to the direct influence of the Atlantic Current have warmed and become saltier. In the Nordic Seas in 1996, warmer Atlantic water, with convection weakened- even though the strength of the current had decreased- increasingly stayed with the current into the Arctic Ocean, contributing to warming and melting there.

An even more intense warming began in August of that year when an anomalously warm water mass flowed into the Iceland Basin, having accumulated in the Gulf Stream over the previous two years. The anomaly traveled westward to the Irminger Sea, reaching the Labrador Sea by 2000, increasing the heat and salt content in the surface layer there by some 12%. The warming trend again intensified in 2001, spanning the area all around southern Greenland, including the Labrador and Irminger Seas (Greiner, 2006). Temperatures also began to increase in Davis Strait and the nearby coastal waters of western Greenland, an area that had been cooling for several decades.

Arctic air temperatures are increasing at a greater rate than most any other area, although the tropics, naturally, remain much warmer. In the summer of 2010, an iceberg measuring 251 km^2 (156 mi^2)- the largest in decades- broke away from Greenland's northwest coast. Glacial melting in all of the coastal areas around Greenland, until more recently, increased at a surprising rate (Greenland Glacier, 2010).

Decline of the Subpolar Gyre, and redistribution of ocean circulation

The abrupt negative shift in the NAO and weakening of the westerly winds in the winter of '96 also initiated large changes in sea level. The dominant trend during the early 1990s was consistent with the generally positive phase in the NAO that had prevailed since the 1980s, with sea-level lower in the Labrador and Irminger Seas and higher in the Gulf Stream and the Atlantic Current. When the climate, or the NAO, turns to a negative phase, the reverse occurs, and sea levels rise in these two seas and become lower in the stream and current. As the climate shifted to a strong negative phase that winter, and continuing for the next two years, the subpolar region, as compared to the levels that prevailed during the early '90s, experienced increases of up to 9 cm (3.5 in) and in the Gulf Stream, the Atlantic, and Azores Currents, levels unexpectedly fell by as much as 15 cm (almost 6in) (Esselborn, 1999).

These sea-level changes are important because they are directly related to the strength of ocean circulation. Increasing sea levels are associated with warmer (expanding) water while decreasing levels mean cooler (contracting) water, reflecting yet another aspect of the generally declining climate conditions.

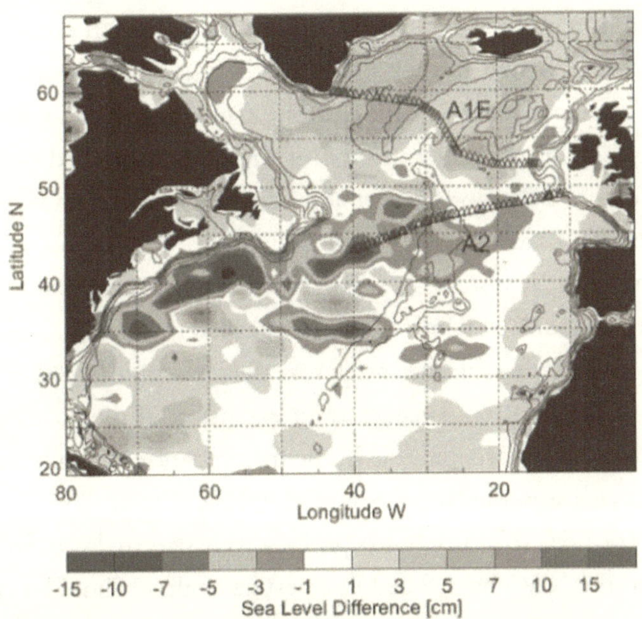

Satellite image of sea level changes in the North Atlantic during the mid-90s. Darker tones lie along the North Atlantic Current, where sea levels decreased, and lighter tone areas to the north denote increasing sea levels—source: US National Oceanic & Atmospheric Administration (NOAA) (Esselborn, 1999).

In the two years from 1996-98, data analysis of surface waters reveals a redistribution of Arctic and Atlantic-derived water masses in the North Atlantic. On the western side of the Atlantic, an atmospheric frontal system- called the Subpolar Front- defines a boundary area between cold and fresh Arctic-derived water in the Labrador Sea region and the warm salty water of tropical origin in the North Atlantic Current. Beginning in 1996, this frontal system, along with the associated Atlantic Storm Tract, shifted slightly eastward, which weakened atmospheric conditions in the Labrador Sea necessary for active heat transfer and convection. Correspondingly, in the northeastern Atlantic, there was a westward shift of Atlantic water in the northern branches of the current- in the Iceland Basin- indicating a contraction of the ocean's circulation system that spanned most of the width of the North Atlantic Basin. Besides this longitudinal contraction, a latitudinal one is evident as well, with a southward shift in the Gulf Stream Core and a slight northward movement of the Azores Current farther south. At the same time, a northwestern shift of Atlantic water from the current itself suggests that within this larger contraction, a smaller version of the same thing had occurred in the area of the Labrador Sea. In addition to the speed and directional changes in the current already mentioned, these events represent a critical weakening of the region's circulation system (Esselborn, 1999).

A very wide and slow-moving current called the North Atlantic Drift circulates from Ireland in the east across the far northeast Atlantic to the south of Iceland. Often considered an extension of the North Atlantic Current, it plays an essential role in the maintenance of a mild European climate, as well as in the North Atlantic, and the climate worldwide. Since the winter of 1996, this current has progressively slowed down. Subsequent changes in the NAO have failed to arrest its decline. Should it continue, this one aspect alone could indicate a reorganization of ocean circulation and change on a global scale. More to the point, this particular change is directly associated in the climate record with the onset of a glacial climate. Now the coldest sea surface temperatures today are more toward the central Atlantic, but a cooling in this region, which we now have with the Blob hard pressed against its western and southern perimeters, is found in the past to have preceded the onset of an ice age by as little as ten years (Satellites et al...., 2004).

3

FRESH WATER & ICE

A late 20th-Century Arctic surge

Over the last several decades, there has been a large, continuing, and increasing flow of Arctic water into the Subpolar Seas and the Atlantic, and this is identified as the immediate causative factor in the suppression of the convection process. The southward-flowing Arctic water is usually fresher and colder than the Atlantic-derived water that it replaces, lowering density contrasts that are crucial for the occurrence of convection. A freshening was observed as early as 1953 in the Subpolar region, and by 1965, there began a sustained and widespread freshening that included the entire Atlantic basin. A significant surge occurred in the late 60s, and by the mid-70s, an almost linear increase in freshwater began in the Nordic Seas, followed again in the mid-90s by another, even more significant surge. Arctic water, both from the Arctic Ocean and glacial melting on Greenland has increasingly flowed into the Nordic Seas and the North Atlantic region every year and has passed equatorward along the western continental margin (Curry, 2005).

The most visible evidence of an increased Arctic water presence is an increase in sea ice; as the volume of water from the Arctic increases, there is a corresponding increase in sea ice and icebergs. Being strongly influenced by the southward flowing East Greenland Current, the waters of the Greenland and Iceland Seas experienced an increasing inflow of sea ice from the Arctic- primarily through Fram Strait- beginning in the

1970s, which peaked in 1995, and the winter season that followed. The flow of fresh water and ice from the Arctic into the Greenland Sea was robust during the early 90s, with almost no local ice formation occurring there. The prevalence of sea ice in the Nordic Seas corresponds with increases of icebergs in the northwest Atlantic beginning in 1972.

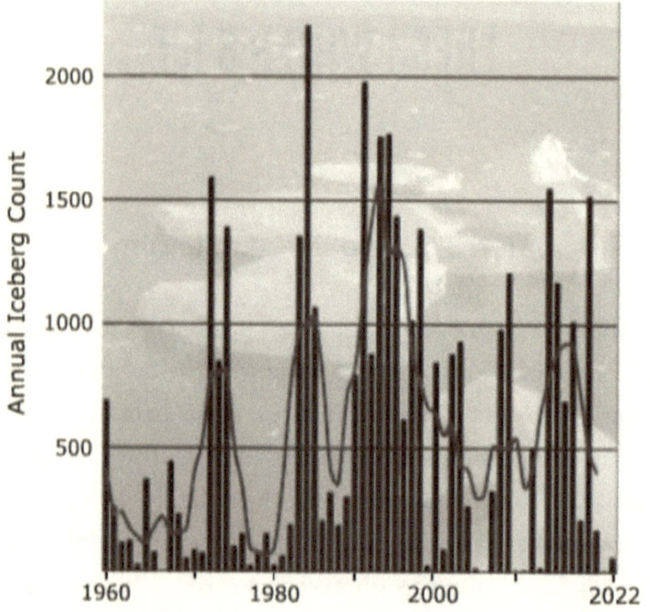

Icebergs south of 48° latitude: This shows the great increase into the northwest Atlantic- especially after 1980- the curved line indicates a five-year average.
Source: US Coast Guard (International et al., 2022).

Off the coast of Newfoundland, in eastern Canada, north of the Grand Banks, and particularly during the spring and early summer thaw season, icebergs number in the tens of thousands, coming primarily from the melting and breaking away (calving) of ice from Greenland's eastern and western coastal glaciers. As the ice flows southward, it tends to melt very quickly as it comes into contact with the warmer waters of the Gulf Stream; icebergs that make it farther south, however, can become a hazard to Atlantic shipping and those flowing south of 48° latitude are watched and recorded- there is a record of iceberg counts in this area going back to

1900. Before 1972, iceberg counts here were fairly low and lower by about half during the 50s and 60s compared to the first half of the century. The annual count of three of the four years of heaviest ice flow from 1900 to 1971 barely exceeded 1000. And for the rest of the century, from 1972 on, the count was higher than that of the previous 71 years combined. The increase in the early 70s was followed by a larger one in the early to mid-80s; then, in 1991, a great surge began that was larger than the 70s and 80s events combined. The most recent increase began in 2015, coincident with the appearance of the Atlantic Cold Blob- detailed in Chapter 8.

The water column in the North Atlantic and the Subpolar Seas, as is true with oceans everywhere, is characterized by a system of distinct layers. Typically, there is a surface layer that comprises the top 500 meters (1,600 ft), an intermediate layer that goes down to about 2000 m (6,500 ft), and a deep layer from there to the bottom. Aside from the mixing in the convection regions, each layer remains largely cohesive with its own distinct temperature and salinity levels. Anomalous water masses that have occasionally had significant impacts in the region exhibit these same properties. One of these, 'The Great Salinity Anomaly,' was a particularly large freshwater mass that was the result of an Arctic surge in the late 60s, and maintained its physical and chemical characteristics for 14 years as it moved through the region in the early 70s (Curry, 2005). The anomaly that began the significant warming of the Subpolar region in 1996 is another important example.

The largest top-to-bottom change
ever measured in the oceans

The 30-year period from 1965 to 1995, in particular, represents the greatest full-depth- top-to-bottom- changes ever measured in the oceans. The entire water column in the Labrador Sea began to cool and freshen in 1965. In 1996, this process reversed in the surface layer, as it began to warm and increase in salinity due to the weakening of convection and resulting accumulation of Atlantic water. The cooling and freshening, nonetheless, continued at depth, and a cold freshwater cap, derived at least partly from coastal melting in Greenland, afterward formed at the very top of the surface layer- an event which greatly interfered with the convection process.

In the Greenland Sea, conditions are different; the water there, especially at depth, is the coldest and densest in the Arctic. The inflowing Arctic water, therefore, has warmed and freshened the surface and intermediate layers there since the late 1960s. In 1972, warmer and more saline Arctic deep water began to flow laterally- directly from the Arctic deep layer- into the deep layer of the Greenland Sea, which intensified by 1980. The Arctic deep water passed downstream through Jan Mayan Channel; this is near Jan Mayan Island in the central Nordic Seas and is the single connecting link for deep water between the Greenland and Norwegian Sea basins. By 1986, this in-flowing Arctic-derived water began to warm the deep layer of the Norwegian Sea significantly; its influence was such that it began pushing upward into the water column, reaching a depth of 1200 m (4,000 ft) by 1990 (Østerhus & Gammelsrød, 1999). Even though the deep layer is becoming more saline, the dominant influence, besides the freshening in the upper layers, is the overall warming, the effect of which is to decrease water density, weakening the convection process there. In the Greenland Sea, the density of the deep layer declined to the extent that by 1994, the density of Norwegian Sea deep water had become greater, and the flow through the Jan Mayan Channel reversed directions, now flowing the other way. This is a significant development because the cold deep current from the Greenland Sea that previously flowed through here made a critical contribution to the density contrasts and the total strength of the greater current flow through the Faroe Channels and, ultimately, the Atlantic Deep Current- that LNADW we have talked about.

The surge that began in 1990 brought large amounts of fresh water and ice into the surface layer of the whole region and beyond, and by the time it finally peaked in the winter of 1995/96, large areas in the high latitudes had been cooled and freshened by the expanding influence of Arctic water. At its peak, the increase in the volume of sea ice was comparable to that which accompanied the 'Great Salinity Anomaly' of the '60s; with the mid-nineties event, however, a much more significant proportion of ice was retained within the Greenland, Iceland, and Norwegian Seas and this has potentially significant implications concerning the course of climate change in the region. A similar period of increased ice flow from the Arctic is known to have preceded the onset of the Little Ice Age in the early 1300s.

The Iceland Sea is a frontal area influenced by water from the Arctic and glacial melting on Greenland, as well as the warmer water from the North Atlantic Current. Much of the water flowing from the Nordic Seas region to become part of the southbound overflow in the Denmark Strait passes through here. The surface waters of the Iceland Sea freshened and cooled dramatically in the winter of 1996- temperatures fell by more than 3° C (5.5° F)- returning to normal the following year. These were the lowest temperatures recorded since measurements began in the 1950s (Olafsson, 2000). It is evident that the events in the Iceland Sea were part of a significant expansion of Arctic water that winter that included all of the Nordic Seas, as well as the farther north Barents Sea, and even the Bering Sea in northern Alaska. This is a very sudden and dramatic change, coming about in that one single winter season.

Changes in the type and extent of surface-water plankton are evidence of this Arctic water expansion. Over the next several years, large blooms of cold water, carbonate, or calcium-type plankton covered many of these areas, having nearly totally replaced the previous warmer water, silica-type, species. The broad extent of these cold water blooms eventually subsided, but in the Iceland Sea, by 1999, it had grown to cover an area the size of Belgium, and researchers found that as much plankton sediment had collected on the sea bottom in a two-week period as in the previous 15 years combined (Ostermann, 2000). The sudden depletion of warm water plankton species in this region comes on the heels of a long period of decline- a couple of thousand years. If this represents a permanent change, then it is, as it has been in the past, a signal of a change to cooler climate conditions.

Loss of deep-water convection
in the Greenland and Labrador Seas

In the Greenland and Labrador Seas, as a result of the convection process, in-flowing Arctic water has descended into the water column. The deepening phase in the Greenland Sea was particularly rapid during the early 1990s, and by the winter of '96, it had penetrated deep into the intermediate layer in both seas. Most importantly, a kind of temperature and salinity barrier- termed 'temperature and stability maxima'- became established at the base of the in-flowing Arctic water in both, which inhibits the vertical mixing of the water layers above and below it (Karstensen, 2005; Khatiwala, 2002). Arctic water layers are characteristically more stratified and do not mix well- as does Atlantic water. And this barrier, which remains strong today, has almost put a lid on the already weakened deep-water (overturning) convection process in the two seas where it has been most important.

This is an event at which we should take pause and reflect upon its importance. It happened deep in the ocean, almost entirely overlooked, with all of us land dwellers busily going about our daily routines. But its occurrence set in motion a series of events that will one day be seen to be perhaps among the most important of all with respect to our changing climate, and that will undoubtedly transform the lives of every single person on this planet. The loss of this ocean process, especially in the Greenland Sea, marks the beginning of the end of the warm climate we have known.

The next important shift in the climate began to unfold in 2010 with a significant and rather sudden decline in the northward flow of the Atlantic Current. Following that, there began a truly invasive replacement of Atlantic water with that of Arctic origin. Together, these two events, occurring mostly in the ocean surface layer in the subpolar region of the North Atlantic, represent an even bigger change geographically and a much more visible event than was the mid-90s shift.

With respect to the maintenance of the Atlantic link in the global system of ocean circulation, these two climate shifts, taken together, represent the most significant advancement of Arctic water south of the Arctic Ocean since the beginning of the Holocene. The relatively warm conditions that have prevailed for most of the last 15,000 years, aside from a 1,000-year period of cold early on, called the Younger Dryas, is absolutely dependent upon a vigorous circulation of warm Atlantic water through the North Atlantic and into the Nordic Seas to the Arctic. The near cessation of deep-water convection and the weakening of the Atlantic Current are attributed to this really invasive southern flow of Arctic water. Together, they should be seen as a signal of a profound and long-term change, and not good news for the continuation of our warm interglacial climate- that has prevailed since before the collective memory of human civilization.

PART TWO

THE PAST AS PROLOGUE

ORBITAL PARAMETERS

The forces of major change

Now that we have looked at some important recent changes in the North Atlantic let's begin to explore the influences that bring about those changes. In this chapter, we'll investigate those events that determine large climate change and perhaps find answers to some fundamental questions, such as why the climate alternates between ice ages and times like today- interglacial periods that are relatively warm. Indeed, why does the climate ever change at all? An understanding of past climate is our keystone, so to speak; it is crucial to understanding present conditions and trends and in projecting what future course the climate may take. The Earth's climate, like many aspects of our environment, including the seasons of the year and our very lives, is cyclical. The familiar cliché, 'the past is prologue,' suggesting that the past will tend to repeat itself, may be at least as applicable to climate change as to the historical events to which it is usually applied. For the last 1.5 million years and more, the climate has been dominated by glacial conditions. Great ice sheets have advanced and retreated many times, on the northern continents in particular, and this is generally understood to be part of a climate cycle governed by long-term changes in several orbital parameters of the Earth.

Popularly called the Milankovitch Cycles, after the Serbian mathematician by the same name, who was one of the first to describe them, the first two of these parameters concern the Earth's axis of rotation. We can imagine ourselves taking up a position far away and above the Northern Hemisphere, where we can have a front-row seat to the goings-on with the entire Earth/Sun system. In that case, we will notice its daily counter-clockwise rotation on its axis, and we will see it moving along the path of its orbit as it makes its annual revolutions around the Sun- in that same counter-clockwise direction. Over the span of a human lifetime, or even over a period of hundreds of years, its axis appears to always point to the same location in the sky. The point toward which the northern polar axis is angled can be found on a clear night near the star Polaris, and in the Southern Hemisphere, the axis points out a location near the star Sigma Octantis.

High summer and the precessional cycle

Our planet's axis is at an angle relative to the plane of its orbit- its ecliptic. And as it moves along through its orbit, that angle changes in relation to the Sun, and this produces the seasonal changes that occur in the middle and high latitudes in both hemispheres. Over much more extended periods of thousands and tens of thousands of years, however, both the angle and direction of the Earth's axis change and this, in turn, produces profound long-term changes in the climate.

The axis itself is found to describe yet another rotation, which occurs over a very long period of time. From the perspective of our front row seats, however, this time around, it is in the opposite- clockwise- direction and at a rate of rotation in sync with that of its revolution. In other words, during this 23,000-year cycle, the Earth's axis makes one single turn, opposite to its annual motion, all the way around- 360°. And having completed that turn, it finds itself at the exact same angle and in the exact same place in its orbit, where it began. This is the first of these orbital parameters and is called 'Precession'- a word to indicate an oppositely rotating effect.

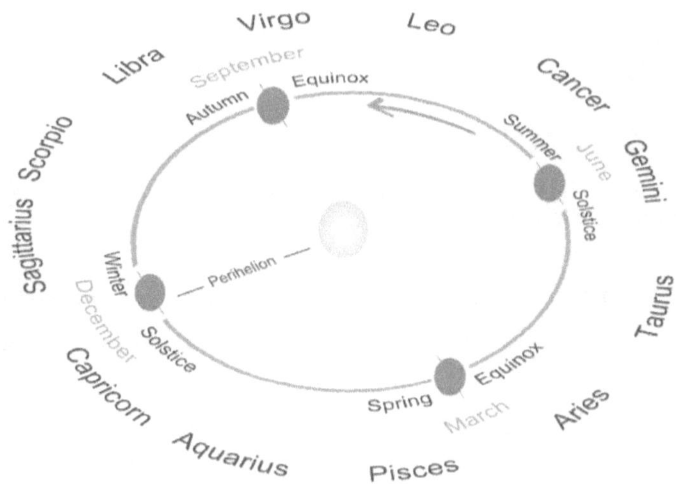

The present configuration of the Earth/Sun system and the precessional effect: The arrow indicates the annual counter-clockwise revolution of the Earth around the Sun, as viewed from this vantage point above the Northern Hemisphere. Its daily rotation on its axis is in the same (west-to-east) direction. Zodiacal constellations are added for points of reference.

Its turn is much slower, but it can be likened to the spin of a gyroscope or a child's toy top. The size of the circle described in this motion varies slightly over even more extended periods, and there is no universal agreement on the time involved. However, today, the 23,000-year time period coincides well with long-term changes in Insolation, as well as, for example, the oppositely located orbital locations of the summer and winter solstice events.

Another condition to add to the mix involves changes in the relative Sun/Earth distance. The Sun is found not to be at the very center of the Earth's orbit. The Earth is about five million kilometers (3 million miles) closer to the Sun on one side than the other. Its point of closest approach is referred to as 'perihelion,' and its opposite, farthest approach-'aphelion.'

Summer Solstice is an important event in the annual warming of the climate; it is, in a real sense, the height of summer, or 'high summer' as it might have been known long ago. It is the time of year when the angle of the Earth's axis is inclined most directly toward the Sun. It is an event that plays a determining role in this first major cycle of climate change.

From its orbital position in the Early Holocene, Summer Solstice today has moved to its opposite (aphelion) point.

At one point in this cycle's 23,000-year journey around the Sun, the Summer Solstice will correspond to the Earth's point of perihelion- its closest approach to the Sun. In other words, there is one point in every precessional cycle when the polar axis is at its most direct angle toward the Sun at the time of closest approach. The conjunction of these two events in the Northern Hemisphere, with its great span of landmasses and greater potential for changes in ice, snow cover, and solar reflectivity, sets the stage for a major climate change. Northern summers receive more direct sunlight and become warmer- a point of truly high summer- a point we will call, for purposes of discussion, precessional maximum.

These are the orbital conditions that prevailed some 11,500 years ago as the present Holocene interglacial period began. It was such a time of high summer- of precessional maximum. Insolation had reached its peak in the northern latitudes, and the entire world was warmed. To the observer in the Northern Hemisphere, the night sky would have looked noticeably different from today. The north polar axis would have pointed out a location near the bright star Vega, and the stars in the nearby constellations of Hercules and Cygnus, because of the daily rotation of the Earth, would have been 'circumpolar stars,' always appearing to circle Vega, its constant companions in the northern sky. Groups of stars now hidden below the southern horizon would have- in their rotational turns- dominated the southern sky, and Polaris- the present north star- along with the present circumpolar stars, such as those in the nearby constellations of Ursa Major and Ursa Minor- the Big and Little Dippers- would have taken their turns high in the night sky as the Earth continued in its daily rotation. The constellation Draco is an exception; between Vega and Polaris, these stars are essentially in the middle of the precessional circle and so are constant circumpolar companions, no matter where the north polar axis happens to be in the precessional cycle.

Continuing to observe the precessional effect upon our return home from our little space adventure and tracing the precessional pathway pointed out by the Earth's axis in the night sky, beginning at Polaris and progressing through an entire cycle, our line will describe a great circle with the constellation Draco directly at its center. From our vantage point here on the Earth's surface, however, we will be seeing its flip side from before, so to speak, so such long-term observations of the stars along the precessional circle will be seen to move in an opposite- counter-clock-wise direction. Ancient Egyptian priests looking up from their pyramids at the beginning of the historical period and the advent of the Old Kingdom some 5,000 years ago would have seen the star Thuban as their north star.

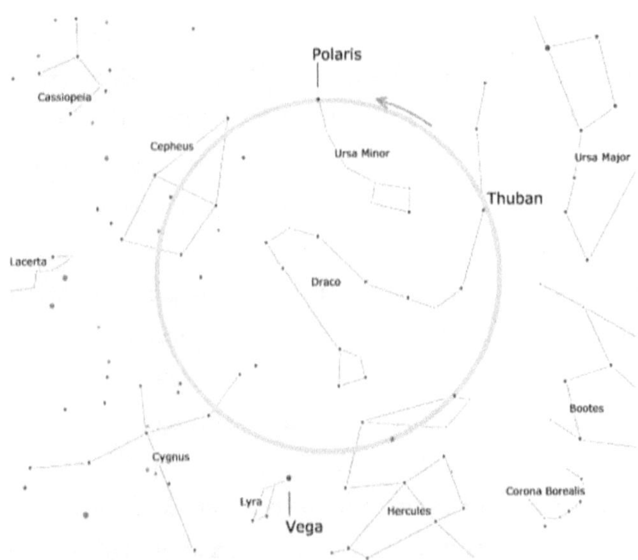

The precessional circle and the changing north star: Note the opposite counter-clockwise view of its motion as seen from the Earth.

Now, the precessional motion of the Earth's axis produces an interesting seasonal effect. The occurrence of seasonality is directly attributable to the angle of its axis relative to the ecliptic- otherwise, there would be no seasons at all. Along with the precessional migration of the axis, there is a close and corresponding change in the seasons- right along with it. A long march, or progression, of the seasons and orbital events such as the Spring and Autumn Equinox and Summer and Winter Solstice, by which we mark their passage through the calendar year. The term has been applied most often to the Spring (or, Vernal) Equinox, which has been important since ancient times as the beginning of the growing season.

The most important of these for our purposes, of course, is Summer Solstice. It is the longest day of the year and the beginning of summer, and in the Northern Hemisphere, it occurs today in late June- usually on the 21st. Those ancient priests, however, working through their astrological and astronomical calculations, as a result of this effect, would have celebrated it in early Autumn- in September on today's Gregorian calendar. Whereas today's calculations would place the Sun in the sign of Cancer at that time of year, our ancient priests would have seen Virgo to be the house of the Sun. In like manner, the Spring Equinox also occurs earlier and earlier in the year. And if future generations wish to continue celebrating Christmas in early winter, for example, it too will require some revisions to the calendar. Left as it is it will eventually become a more wintry holiday.

The progression of the seasons through the calendar year, in reverse order to the flow of time, has been observed over a long period- since antiquity. Every sixty-two years or so, the seasons and the Solstice and Equinox events, along with them, occur about a day earlier. It is an effect known as the 'Precession of the Equinox,'

Today, our tracing of the path of the northern axis since the beginning of the Holocene reveals it to have progressed about halfway through its precessional circle. Now, on the opposite side of its cycle, pointed toward Polaris, precession is at its minimum in the north. Winter Solstice, which marks the beginning of the winter season, is the shortest day of the year and now occurs in the north on or about December 21st. Very near the Earth's point of perihelion- its closest approach to the Sun, which now occurs on January 4th- an important indicator of where we are today in the climate cycle. The point of absolute minimum in the current precessional cycle has, therefore, already occurred for the Northern Hemisphere this time around. This happened some 900 years ago as the Winter Solstice occurred at the precise point of perihelion. And so the early Holocene high summer in the north is now long since past. As there is an alternation of the precessional effect between the two hemispheres, it is now the turn of the Southern Hemisphere, at the same time having just passed its absolute point of Summer Solstice and corresponding maximum in this cycle- which we will refer to as Southern Precessional Maximum (SPM)- to receive its season of relative warmth and light.

Axial Inclination and Orbital Eccentricity- more conditions for a climate warming

A second condition, which is just as essential to produce an interglacial warming, concerns the inclination, or tilt, of the Earth's axis. Over a period of approximately 41,000 years the angle of the Earth's axis in relation to its plane of orbit changes from 21.2 to 24.5 degrees, and back again. Just as previously described with the long-term change in the size of the precessional circle, the very same effect of its slight widening and contracting again over the period of this cycle is evident as well. When the Earth is at the point in this cycle that produces the maximum 24.5° angle- a point we will refer to as Axial Inclination Maximum (AIM)- and it occurs in conjunction with a precessional maximum in the north- we will call it Northern Precessional Maximum (NPM)- the angle of the northern axis is tilted even further toward the Sun, and the amount of direct sunlight, and Insolation, in the northern latitudes is at its absolute greatest extent- just enough, it appears, to produce an opportunity for a warming. At its maximum extent in this cycle about 7,000 years ago, the Sun would have been just slightly higher in the sky- shifted farther north- during all seasons as compared to today. The inclination of the Earth's axis has since declined about one degree so that its angle now is at about 23.5°.

The third and final determinant is Orbital Eccentricity or the shape of the Earth's orbit. Today, the orbit is nearly circular, but it varies from nearly circular to a more elliptical one in a cycle of about 100,000 years. An interglacial warming does not have to occur at a specific point, but the range of solar energy received at the surface may vary too widely when the orbit is more elliptical; the range may be some 20-30% at its greatest extent. A warming only occurs in the part of this cycle when the Earth's orbit is more nearly round (McBean, 2004; Milankovitch Variations, 2006; Astronomical Theory, 2009; Milankovitch, 2011).

It appears, from the results of this study, that when the maximum points in the precessional and axial inclination cycles occur within a few thousand years of one another in the Northern Hemisphere, and this happens when the Earth's orbit is more nearly round, the climate warms up to interglacial conditions, like those of today. The indications are that these periods of warmth occur only every 70,000 to 100,000 years or so and may last for 10,000 to 15,000 years-and sometimes more. This analysis indicates that for at least the last 1.5 million years, the prevalent cycle for interglacial periods has been about 82,000 years from the onset of one to the onset of the next. The most recent maximum in axial inclination, 7,000 years ago, occurred some 4,500 years after a precessional maximum in the Northern Hemisphere that marked the beginning of the present Holocene interglacial period, and it is no coincidence that the warmest and most humid climate of this interglacial period occurred between these two points.

As we look back over time at previous interglacials, each one is a little different, but there are also similarities and some interesting correlations. We find, in this analysis, that an alternating progression of the first two of these same orbital conditions defines the beginning and early part of each period of warmth. For example, the most recent interglacial warming, prior to the Holocene, was the Eemian Period, which began 130,000 BP (years before the present) when the Earth's axis was at its maximum in the inclination cycle. A precessional maximum occurred in the north several thousand years later, at about 126,500 BP- in reverse order to the Holocene. During the next most recent warming- the Penultimate Interglacial- beginning at about 218,500 BP, the timing was different, but these two climate events occurred at that time in the same order as the Holocene. Both the Eemian and the present Holocene, besides being generally warmest during the early part of the period, are also characterized by a mid-period of optimal warmth before turning to a later period of cooler, dryer climate. Temperatures during the early Eemian, however, were several degrees warmer than today; sea levels were higher, and Greenland was greener- an effect of its AIM and NPM occurring closer together. Orbital Eccentricity was more, too, resulting in generally greater seasonality, with wider extremes in summer and winter conditions (Adams, 1999). The earlier Penultimate Period, on the other hand, was cooler and never reached temperatures as warm as the Holocene.

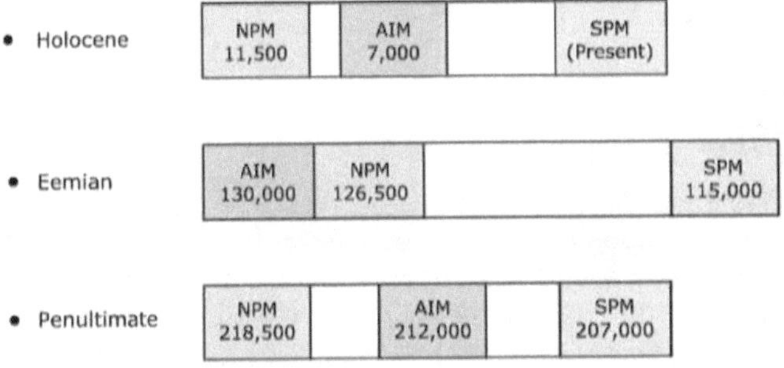

The sequence of climate events for the present and two most recent interglacial periods: Numbers are in years before the present (BP).

The coming together of these orbital parameters is critical to the occurrence of an interglacial warming. The sequence in which they occur within an interglacial period plays a determining role in how the warming progresses and in its duration, although the evidence also shows that climate events are not strictly confined within the limits of their occurrence. The warmer temperatures of the Early Eemian, for instance, were characterized by an Axial Inclination and Northern Precessional Maximum that were a thousand years closer together than occurred during the early Holocene, and the parameters that define the cooler Penultimate were more evenly spaced, and farther apart. Now, some 11,500 years since the Holocene began, we find the climate today at its point of Southern Precessional Maximum. At this point during the Eemian, after 15,000 years of interglacial warming, the climate returned to ice age conditions.

The Penultimate, however, endured for some 20,000 years; temperatures were about 2° C (3.5° F) cooler than today, and the winters would have been noticeably colder. Nevertheless, it lasted some 8,000 years beyond its point of Southern Precessional Maximum before full ice-age conditions returned.

Temperature Change (°C)		AIM	NPM	SPM	Orbital Eccentricity
-8 -6 -4 -2 0 2	Holocene	7,000	11,500	Present	Minimum
	Interstadial (1)	48,000	57,500		Maximum
	Interstadial (2)	89,000	80,500		Near Min.
	Eemian	130,000	126,500	115,000	Near Min.
	Interstadial (3)	171,000	172,500		Maximum
	Penultimate	212,000	218,500	207,000	Minimum

Comparison chart of events for the last 212,000 years: The temperature scale is a comparison to the present-indicated by "0." Interstadials are lesser warm events when all three climate parameters are not in conjunction. The NOAA is the source for the graphic data, derived from the Vostok Ice Core- the longest continuous temperature record (Petit, 1999).

Projecting these parameters into the future, using what we have learned about the major forces of climate change, and assuming that they will hold for the duration of the next glacial cycle, we find that their next conjunction will not occur until about the year 77,000 CE (Common Era- same as AD). We can expect this future warming to begin with a maximum in the Earth's axial inclination cycle, as what occurred at the beginning of the Eemian, and that it will be followed some 5,500 years later by a Northern Precessional Maximum, a sequence in the opposite order, and about 1,000 years longer, than the Early Holocene, suggesting that it may be of a similar over-all duration.

Analysis of these orbital conditions and their determining influence on the Earth's cycle of glacial/interglacial climate should impress upon us the inevitability that is inherent in large climate change. Our most relevant lesson from this is that today, we live toward the end of an interglacial period, and the conjunction of the orbital parameters that produced it will not come together again for a very long time. We can expect that the great ice sheets that melted away as the warming of the Holocene began will one day begin to grow again and that ice age conditions will return. We do not know precisely when that will occur. However, a look at the history of climate events over the course of the Holocene, and in particular, some more recent and dramatic changes that have occurred in the northern seas, will help us narrow our focus on how those final days of the Holocene might actually play out.

5

THE CLIMATE RECORD

Getting a Perspective on the Past

I t is said that you cannot know where you are going unless you know where you have been. It is the record of the past, the paleontological or climate record, that is the most valuable tool at our disposal as we endeavor to understand the changes we see going on today in the North Atlantic and globally. Analysis of core samples taken from sediment layers in glacial ice, deep-sea beds and lakes, and other sources, such as tree-ring data, provide a history of climate that extends far into the past.

A little over 50 million years ago, a peak in a period of warming occurred- called the Eocene Optimum- after which the Earth's climate has progressively cooled down by at least several degrees. For reference purposes, this would have been almost 15 million years following the Cretaceous-Tertiary Extinction Event. Unrelated to the later Eocene warming, it was caused by the impact of a giant asteroid, 65 million BP, an important event in Earth's geological history; it may have killed 70% of the life on Earth at the time, including the dinosaurs. At any rate, by about 33 million years ago, the East Antarctic Ice Sheet had begun to expand across the continent, and by 11 million BP, the ice had reached its western side. Today, nearly all of Antarctica is ice-covered, including large spans of coastal water areas where great ice shelves have formed. In recent decades, some of these same ice shelves have been in the news; large portions have, in some instances, broken away from the continent as the climate has warmed- especially along the coastal areas of the western peninsula. The overall cooling has become more rapid in the last 15 million years, and conditions in the Northern Hemisphere have progressively become more conducive to ice formation. Essentially, the Earth is amid a great ice age that began over 30 million years ago. Nothing like it has occurred in recent geological time; the most recent comparable period of glaciation ended about 260 million BP and had endured for some 60 million years.

Some 3 to 3.5 million years ago, glaciation began in earnest in the north, starting with Greenland and extending to the northern reaches of Europe and Asia and to northeastern North America by about 2.5 million years ago (McBean, 2004, pp. 46-47). In the last 2 million years, the climate has continued to cool, and ice sheets have progressively expanded over continental land areas. For the last 1.6 million years- comprising the period of the Pleistocene Ice Age- the climate has oscillated between long periods of ice age conditions and relatively short intervals of interglacial warmth. The Atlantic Ocean did not exist before about 200 million BP when an ancient supercontinent called Pangaea began to break up. Due to plate tectonics- the movement of the Earth's crust over time- there is a very slow, primarily east-to-west, drifting of the continents away from one another and, consequently, a progressive widening of the Atlantic Basin. Pangaea included all of the Earth's landmasses and formed from an even more ancient drifting together of the continents; so, the Atlantic Ocean is a result of this more recent continental drift. A mid-ocean rift system spans the North and South Atlantic, where the seafloor gradually spreads apart as part of this process; volcanic activity occurs along the rift as new seafloor is created. Iceland, too, has formed as the result of this same volcanic process; it lies directly on the rift and is at the very center of the area of the Subpolar Seas, which is the primary area of concern in this analysis.

Following the end of the Eemian Interglacial Period, which had begun some 15,000 years earlier, ice age conditions persisted for about 100,000 years. The climate during this time oscillated between cold and colder conditions over decades to millennia. The Nordic Seas were, most of that time, completely ice-covered, and in the coldest times, winter sea ice extended into the Atlantic as far south as present-day Britain. Then, almost 15,000 years ago, things began to warm up; sea ice and the vast continental glaciers began to melt, and Atlantic water began to flow into the Nordic Seas through an ice-free corridor that opened up along the Norwegian coastal area; this was along the continental margin, where the shelf area ends, and the contour of the sea bottom drops off toward the deep ocean floor. Called the Bolling- Allerod Period, the warming was interrupted about 12,500 BP by a sudden return to ice-age conditions- the Younger Dryas. Geological evidence shows the Dryas flower to have extended, along with the primarily treeless tundra, into the middle latitudes during glacial times. The Younger Dryas was the most recent of large cooling events; the near glacial conditions that prevailed immediately prior to the Bolling-Allerod warming and the full-blown ice age temperatures that prevailed before then, being sometimes referred to respectively as the 'Older' and 'Oldest Dryas.'

The Younger Dryas event was brought on by a release of freshwater into the Atlantic from a large lake called Agassiz that had formed in what is today south-central Canada from melting glaciers. From the very onset of interglacial warming, there was a great influence of relatively fresh water from the Arctic, and from the large continental-sized ice sheets as they slowly melted away. The melting contributed significantly to a freshening of the Atlantic for several thousand years- until about 7,500 BP- with melting from the Hudson Bay region lasting a little longer. Lake Agassiz may have become as large as the present Hudson Bay and Great Lakes combined. Initially, the accumulating water from this lake drained through the Mississippi Valley, but eventually, melting ice farther east allowed a rather sudden flood of water to drain through the Great Lakes region and St. Lawrence Seaway and on into the North Atlantic. Other similar, although smaller, glacial lakes formed to the west and south of Agassiz; Lake Bonneville, which formed in what is now the northwestern United States, is one example. All of them produced significant erosion downstream, carving out riverbeds, valleys, and canyons along the way. The climate record shows that at the advent of the Younger Dryas, ocean circulation and convection in the climate-sensitive North Atlantic was severely weakened by the resulting freshening, and ice age conditions returned for about 1,000 years (Broecker, 1999).

The source of an increased presence of freshwater in the Atlantic region is not usually so dramatic as the drainage of a large glacial lake. A certain amount of it is expected as normal, flowing south every year through the East Greenland Current and other usual channels from the Arctic- primarily through the Nordic Seas. Such an event is particularly relevant to us today, with the widespread freshening occurring now in the Atlantic and the northern seas, because every major cooling event in the North Atlantic region has been accompanied by such an invasive flow of Arctic-derived water.

The Early Holocene: A time of truly northern summer

When the climate finally warmed up again, at the beginning of the Holocene, it did so fairly rapidly and simultaneously in all regions of the world; temperatures as warm as today were not achieved until about 10,000 BP, but the initial warming was the most rapid of major climate events on record. For people living at the time, a mid-summer celebration would have had an added meaning. Well within the span of a human lifetime, the climate emerged from the cold conditions of the Younger Dryas to the warmth of a truly Northern Hemisphere summer. The northern oceans experienced a pronounced warming, with a strengthening of the flow of Atlantic water throughout the North Atlantic region and the Subpolar Seas. The Early Holocene, from about 11,500 to 5,000 BP, was a period of 'Climate Optimum;' summer Insolation was very high, exceeding present values by 8 to 10% in the North Atlantic (Drange, 2005). Climate conditions were generally strong, with warmer temperatures and higher humidity, and a strongly positive trend in the NAO, certainly stronger than we have seen in modern times. The one exception was a 200-year-long temperature drop 8,200 years ago, which, like the Younger Dryas, is attributed to Arctic melting and an increased southward presence of fresh Arctic water (deMenocal, 2000). By 7,000 BP, northern glaciers had retreated to positions behind today's limits, and many were completely melted. Sea levels had risen some 34 meters (112 ft) above that of the Younger Dryas and perhaps 100 m (325 ft) above the levels that prevailed during glacial times. All coastlines worldwide were inundated by the rising seas, having by this time achieved levels in many areas approaching those of today's. During this early period of 'Climate Optimum,' northern forests expanded to their maximum Holocene range, where temperatures were at least several degrees warmer than at present (McBean, 2004).

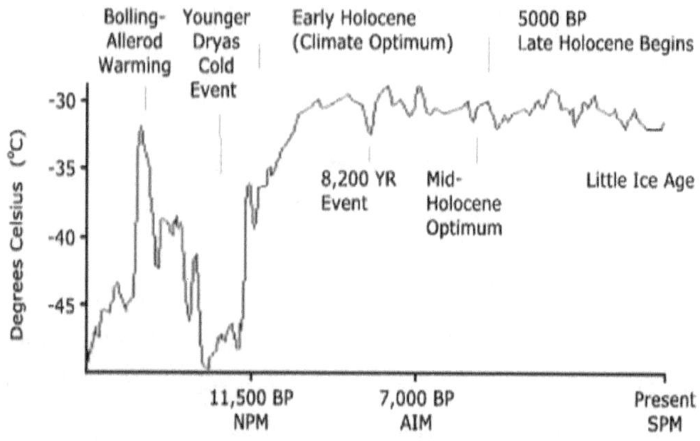

History of Holocene temperature change: Climate events and air temperatures in central Greenland for the last 16,000 years. The progression of climate change varies by region; while a mid-period optimum occurred in Europe and elsewhere, it is not very evident in this data for Greenland. The NOAA is the source for the graphic data-collected from the GISP II Ice Core. Each data point represents ½ of a degree change or greater, and each point is within about 25 years of its sample location in the core (Alley, 2004).

As the climate changes back and forth between cold and warm periods, corresponding changes occur in the relative influence of Arctic and Atlantic water. In the Nordic Seas, the Polar Front is an atmospheric system that defines the boundary between these opposing water masses, and the history of the Early Holocene can be defined by its progression from the far south-eastern Norwegian Sea, where it was located at the end of the Younger Dryas, northwestward across the Nordic Seas as the climate warmed. The influence of Arctic and Atlantic water was characterized by some degree of variability in the Early Holocene, at times changing to a transitional pattern where both water masses had a significant presence. By 9,000 BP, the Polar Front and accompanying sea ice margin, or extent, had migrated to positions in the western Nordic Seas, and the presence of Atlantic water significantly increased. And by 7,000 BP- the point of Axial Inclination Maximum- both had retreated to positions along the northeastern Greenland coast, and the influence of Atlantic water in the Nordic Seas had expanded to its greatest Holocene extent- while sea ice extent was at its absolute minimum (Koc, 2000).

About 6,000 BP, conditions began to change; it was the beginning of the largest warming event of the interglacial period- the 'Mid-Holocene Optimum'- which included all of the Northern Hemisphere and tropical latitudes, and lasted for some 1,000 years, producing relatively high sea surface temperatures (SST) in both the tropical and North Atlantic oceans. Even as the climate warmed, however, signs of a cooling began; at about the same time, a transition occurred in the North Atlantic from a predominantly positive North Atlantic Oscillation (NAO) to a more negative one. In Scandinavia, where the warming began and ended earlier than in regions farther south, an abrupt cooling occurred at about 5,800 BP, and forests there began to retreat (McBean, 2004). At about the same time, there were decreasing temperatures and tree growth in Siberia, and tree species in Europe and North America began to decline. The period is characterized by unstable conditions in the North Atlantic and Subpolar Seas; an increased presence of freshwater is evident in the North Atlantic, and a weakening occurred in the Deep Water Current, which indicates a weakening of the surface current and in the convective regions as well. And although this was the high point of the Holocene, temperature-wise, Greenland experienced more winter-like conditions (deMenocal, 2000).

The optimum had ended by 5,000 BP when a significant and widespread downward trend began. On the positive side, the indications are that there was a resumption of convection at this time in the Labrador Sea, where it had been suppressed- perhaps totally so- due to glacial melting since the beginning of the Bolling-Allerod warning, providing a bit of a stabilizing influence to otherwise deteriorating climate conditions (Cottet-Puinel, 2004).

The Late Holocene: a new period of climate deterioration begins

SST began to decline, and a major cold event occurred at 4,800 BP that endured for several hundred years. Glaciers began to advance in northern Iceland, Greenland, and northern Eurasia. The Polar Front, by this time, had migrated to a position north of Iceland, associated with increases in the volume of Arctic water, precipitation, and storm activity in the Nordic Seas. As these changes began in the north, the onset of much drier conditions occurred in the subtropical regions. When the climate turns colder, it also becomes drier, a correspondence that is evident during every period of climate cooling. And the drying effect extends well beyond the northern latitudes; during glacial periods, drought conditions prevail globally. The Early Holocene was a time of high humidity in North Africa and the adjacent Sahel; the region was covered with grassland vegetation, with many small lakes and streams. Around 5,700 BP, however, desert conditions began to expand across North Africa. There was a robust arid event that spread from there into southern Asia beginning about 4,000 BP, which lasted for 400 years, and today, the climate across North Africa more resembles the conditions that prevailed there during the last glacial period (Adams, 1999; Claussen, 1999).

More large cooling events occurred at 3,800 and 2,600 BP, the second of which was particularly severe in northern Scandinavia, with many glaciers there advancing to their maximum Holocene positions. Atlantic water, by this time, had become much diminished in the region, and climate conditions in general were becoming colder and increasingly variable. Compared to 1,000 and 2,000 years ago, SST in the northeast Atlantic and the Nordic Seas and air temperatures in Greenland have cooled by as much as 2° C (Eiriksson, 2006). The change was not gradual, however. Temperatures and climate conditions in general in this region have experienced increasingly large and high-frequency changes. In northern Iceland, six periods of glaciation have occurred since the beginning of the Late Holocene, four of which correlate with ice rafting events in the North Atlantic (McBean, 2004). Ice rafting events are periods of increased volume of sea ice and icebergs, and they are most characteristic of glacial climate, during which massive amounts of Arctic ice may be discharged into the North Atlantic and are associated with the very coldest of temperatures. More recently, these events have occurred with increasing frequency, with Icelandic peaks of glacial advancement at intervals of about 400 to 650 years; a similar periodicity of cooling events has occurred in the far northern and northeastern Atlantic (Witak, 2005).

Change in the last one thousand years

A pattern of warm events followed by periods of cold has become increasingly evident since the beginning of the Common Era, about 2,000 years ago. In most areas, there is a trend toward cooler temperatures that has been more pronounced in the last 1,000 years. The beginning and ending dates vary from place to place, but in Europe, there was a period of warming, the 'Roman Warm Period,' from about 50 BCE (Before the Common Era) to about 400 CE and a later 'Medieval Warm Period' from about 750 to 1425. Both were followed by rather abrupt cold events. The first, called the 'Dark Age Cold Period,' occupied the 350 years between these two warmings, and the beginning of the cold event that ended the second warm period, the Little Ice Age, was particularly rapid and strong. Its onset was preceded by a century of ice-rafting (more sea ice and icebergs), with SST in the Norwegian Sea decreasing by about 1° C (almost 2° F) in a few decades (Eiriksson, 2006). The cold endured for 500 years, exhibiting the most severe climate conditions of the Holocene. Glaciers on Iceland, Greenland, and Jan Mayan Island reached their greatest extent between 1750 and 1850, after which all began to retreat. The absolute coldest temperatures of the Holocene occurred in northeastern Canada between 100 and 300 years ago, only just ending with the advent of 20th-century warming (McBean, 2004).

Since the early 1900s, the sea ice extent has, for the most part, retreated during all seasons. The melting is especially evident in the eastern Arctic. This is particularly so on the European/Siberian side of the Arctic Ocean, where temperatures are increasing, as is the influence of warmer Atlantic-derived water. By late summer in the western Nordic Seas, the ice has retreated all the way to the coast along central and southern Greenland, with a corresponding increase in the melting of coastal glaciers. In winter, the extent of sea ice in this region has declined by about one-third since the coldest times of the Little Ice Age- just a few hundred years ago. In the 20th Century, the retreat has been most evident in the eastern Arctic and a little more so in the western Nordic Seas- but the winter change has not been quite as dramatic as the summer melting.

Looking at this from a somewhat broader perspective, however, the sea ice extent, as well as the presence of Arctic water, is in fact, found to have significantly advanced in the western Nordic Seas as compared to its 3,000 BP boundaries. In most winters, sea ice extends to Svalbard in the north, about halfway to Jan Mayan in the central region, halfway across the Denmark Strait and along the coast of southeast Greenland, and in some winters to the region directly north of Iceland. And while reduced from its more extensive reach during the Little Ice Age, when we consider the bigger picture of climate change, we find winter sea ice to otherwise be at its greatest extent in at least 9,000 years.

Just as the Polar Front migrated westward across the Nordic Seas during the early Holocene, it has, since the point of Axial Inclination Maximum, 7,000 years ago, moved back the other way. The influence of Arctic water is much increased since then. By 3,000 BP, the water balance began to favor that of Arctic origin, with unmixed Atlantic water much reduced. Today, the Polar Front lies in a diagonal position across the middle of the Nordic Seas, its southern portion extending to the northeast of Iceland. The Atlantic Current now mixes with Norwegian Sea water in the area of the northeast Atlantic as it continues in its northward trek into the Nordic Seas and on to the Arctic. This is an important milestone in the evolution of the climate; temperatures in the Atlantic Current have increased in recent decades, but its strength has declined. And, perhaps most importantly, there is no longer a direct influence of Atlantic water in the Nordic Seas (Koc, 2000).

What we have with all this is a mixed bag of warming and cooling. Generally speaking, in the more immediate term, two periods of Twentieth Century warming are evident, with a shorter cooling spell in between and the second one now extending somewhat into the Twenty-First Century. Looking at it all from a broader point of view, on the scale of thousands of years, it is evident that the climate continues a long-term decline. An age-old contest for dominance between Atlantic and Arctic water continues in the North Atlantic and northern seas, with Atlantic water having now lost much of what it gained in the Early Holocene. As we will later see, with the events beginning in 2010, Arctic water has achieved a major new expansion, with associated signs that this second period of our modern warming is coming to an end.

6

OUR PLACE IN THE GLACIAL CYCLE

After five thousand years of cooling

So now it should be pretty evident that the accelerated warming that we have become aware of in the last several decades, except for a small amount of mid-century cooling, is really a century old, and, perhaps surprisingly, it is also quite evident that this warming occurs against a background of a more considerable change toward cooler temperatures that began thousands of years before. In this chapter, we'll take a closer look at several aspects of change that are particularly relevant to where we are in the climate cycle. It is helpful to see how climate events during the Holocene compare with events during the Eemian Interglacial Period and the period of glaciation that followed. Our perspective is further sharpened with an analysis of the cycle of change in the water balance in the northern seas, and we will take a closer look at the period of 'Neo' (meaning new) glaciation that occurred in the Late Holocene, as well as the determining influence of changes in Insolation.

The warm interglacial climate that we enjoy today was preceded by what we consider a very long period, a thousand centuries, of ice age conditions. Analysis of Greenland glacial deposits reveals that during this period, the climate underwent frequent, massive, and abrupt shifts; it leaped back and forth between states of intermediate and extreme cold- not unlike other periods of glacial climate. There were 24 warming episodes, or interstitial events, beginning at the termination of the Eemian Period about 115,000 years ago and ending with the onset of the Holocene. The climate would warm up a few degrees during these warm events, but they never reached temperatures nearly as warm as today's. These interstadial events would begin in decades or less, and their duration might be just decades, or they could last for centuries or even a couple of thousand years. They would end in a series of smaller cooling events culminating in a fairly large cooling, and the climate returned to colder glacial conditions, called stadials. The Bolling-Allerod warming, when the climate first emerged from Pleistocene ice age conditions, is sometimes considered an interstadial event. It was warmer than the ice age stadial events that preceded it but was still a few degrees cooler than the later, more rapid Holocene warming (Adams, 1999).

It is interesting to note that there is an apparent rhythm, or cycle, that the climate tended to follow during glacial times, as temperatures oscillated back and forth. Its periodicity was approximately 1500 years (Adams, 1999). This cycle's origin has not precisely been decided upon, but its timing is a good match with one of several suspected cycles of solar activity. An analysis of major cooling events indicates that a fainter, less apparent effect of this cycle may have persisted into the Holocene. The 8,200-year event fits as one of the more extreme cool periods in this rhythm; aside from the Little Ice Age, which was a much bigger event, it is the most pronounced such episode of the Holocene. Other cooling events correspond to the periodicity of this cycle. The change at the end of the Climate Optimum 5,000 years ago is the most prominent example. The Little Ice Age itself fits as the most recent of this cycle's cooling events.

The Eemian Period- for comparison

A look at the Eemian Interglacial Period provides a good comparison for understanding the climate of our own Holocene. The most recent of past major warming events, it is better preserved in the fossil record. And while there are some differences, the progression of major climate events during that period was very similar to the Holocene. Thus, the climate of the Eemian, especially the latter part, and the termination events in particular, provide insight into how the remainder of the Holocene Period will likely progress and finally end.

The Eemian, in general, due to the Earth having a slightly more elliptical orbit at that time, was characterized by a greater degree of seasonal variability (seasonality). Overall, it experienced a reduced ice volume as compared to the Holocene. Summer temperatures were warmer, and early in the period, winters were cold but soon changed to conditions warmer than today. The early parts of both were times of optimum warmth and similar duration. Both began with two stages of rapid warming. The Eemian, beginning at about 130,000 BP, may have started earlier in temperate and tropical regions. The initial warming of the Holocene, however, occurred more quickly, consisting of a series of warming steps, each lasting less than five years, with about half of the warming coming on in less than 15 years.

Vegetation during the Holocene was present in southern Greenland until some 1,000 years ago. The Eemian, however, was likely greener, and the Greenland Ice Sheet then was less extensive, especially at the mid-period Optimum when the sea level was about 6 m (20 ft) higher than it is now. The Eemian mid-period was a time of maximum ice retreat and enhanced seasonality. Warmer tropics and colder high latitudes resulted in increased precipitation and an intensified hydrological cycle that would have encouraged the development of sea ice in winter and enhanced the southward flow of Arctic water. Thus providing a cold freshwater 'lid, or cap,' which prevented the occurrence of convection in the Labrador Sea.

There is evidence that the Eemian experienced many short duration- a few thousand years- cold events, beginning and ending in decades or less. Evidence from sea cores from the Nordic Seas and the northeast Atlantic reveals a mid-Eemian cooling that had a pronounced regional effect and perhaps was global in reach. Sea surface temperatures declined several degrees; there was freshening of ocean water, as well as changes in ocean circulation patterns. On land, changes in west European vegetation are evident at about 122,000 BP. The cooling came on in a few decades or less and lasted for several hundred years. Afterward, the climate recovered, but conditions did not return to the warmth of the earlier Optimum. The mid-Holocene cooling event, from 4,500 to 4,800 years ago, was more mild but otherwise similar.

In the later Eemian, seasonality decreased- this was 125,000 to 116,000 years ago. Similar to the Holocene, as time went on, the climate system weakened, and summers became shorter and cooler. Eventually, a point was reached where colder winters produced more ice and snow than could melt during summer Insolation, and the Eemian came to a close- giving way to glacial conditions.

The first abrupt cooling event, signaling the beginning of the end of the Eemian, occurred at 117,000 to 118,000 BP. The fossil record of the end of the period is not yet clearly defined. Tectonic disturbances are evident in Greenland glacial deposits older than 110,000 years, and the sequence layer containing the record of the transition event to glaciation is tilted and mixed. It is not yet clear whether this occurred during or after the actual transition. The end event is known to have happened over not more than several centuries, and perhaps considerably less; it is defined by a period of rapid cooling and an abrupt change in deep-water circulation associated with an invasive flow of freshwater into the Norwegian Sea and a reduction in the North Atlantic Current- and these events are linked to the onset of widespread glaciation. There is evidence that a large and explosive volcanic eruption may have been the event that finally initiated the transition to a glacial climate. Ice age conditions appear to have prevailed in the far north by 115,000 BP (Adams, 1999).

An intensifying hydrological cycle, which is playing such an important role now, in the late Holocene, was more active during the entire span of the Eemian. Those determining orbital conditions were a little less ideal then than at present. Like the Eemian, a freshwater cap in the Labrador Sea was also in place during our own early period of warming- and which, incidentally, has recently reappeared. Also like the Eemian, deep-water circulation has today reached a crucial impasse, especially beginning in the mid-nineties. And since 2010, there is a significant weakening in the Atlantic Current, and a large dramatic new presence of Arctic water in the Atlantic. The late Holocene has experienced several warm periods-we may now be at the peak of the present one. All of the previous ones were followed by a rapid cooling; the most recent being the Little Ice Age- which was the coldest its been since the beginning of the Holocene.

Atlantic/Arctic water balance

The relative balance of Atlantic and Arctic-derived water masses in the Nordic Seas and the North Atlantic is an important indicator of our location in the glacial cycle. As outlined in Chapter 5, the Polar Front is the dividing line between these two opposing water masses. Its progression through the Nordic Seas over the course of the Holocene is a good long-term indicator of where we are now in the cycle of change. Although temperatures like today's prevailed by about 10,000 BP, the climate continued to warm up, and glaciers continued to retreat for the next several thousand years. Cold-water plankton in the Nordic Seas was replaced with warm-water species, and the warmth of the Early Holocene Climate Optimum was in full swing. A cold-dominated ocean circulation changed to a transitional pattern, influenced by Arctic and Atlantic water masses, with an associated northwest retreat of the winter sea ice extent-or sea ice margin- and the Polar Front. For the duration of the Climate Optimum, the relative influence of Arctic and Atlantic water masses in the Subpolar Seas and the North Atlantic fluctuated several times. The relatively fresh Arctic water exerted a more decisive influence when atmospheric circulation was weak, when the NAO was in a cold (negative) phase, and the North Atlantic Current was reduced.

Conversely, the warmer and more saline Atlantic water was dominant when the NAO was in a positive (warm) phase, and the influence of the Atlantic Current was strong. At times, both Arctic and Atlantic water had a significant presence in the Subpolar Seas, but the climate, overall, during the early Holocene was characterized by the warmest and most humid conditions of the period. For the last 2,000 years or so of the Optimum, the period from about 5,000 to 7,000 BP, major circulation changes occurred in the region, and the last 1,000 years was a period of significant warming-the 'Mid-Holocene Optimum.' As the climate began to deteriorate, and the new period of 'Neo-glaciation' began 5,000 years ago, temperatures began to decline, and the trend in the NAO changed to a more negative phase. A trend that has continued up until modern times (McBean, 2004).

The Polar Front, during glacial times, is thought to extend southward into the northeast Atlantic; at the time of the Last Glacial Maximum (LGM), some 20,000 BP, it would have reached southward to about 50° North Latitude. All of the Nordic Seas then was ice-covered, and winter sea ice was widespread in the most northern reaches of the Atlantic, whereas summers there, on the other hand, would have been generally ice-free. The type and the relative abundance of surface water plankton that lived at the time, the remains of which settled to the bottom, are found in the layered deposits of deep-sea cores and reveal ancient sea-surface conditions, such as temperature, salinity and ice coverage. As the climate changes back and forth between glacial and interglacial conditions, the Polar Front migrates back and forth (from southeast to northwest) across the region- but its progression is not necessarily a smooth one, nor is it gradual. The evidence from deep-sea cores shows that by the time of the Bolling-Allerod warming, the front had migrated to the area of the North Atlantic near Iceland. A corridor had opened up along the Norwegian coastal margin, extending from the Atlantic to the Barents Sea, but aside from this, the Nordic Seas remained ice-covered. The northern portion of the corridor refroze as the cold of the Younger Dryas set in, and the Polar Front migrated back southeastward to a position at about 60° North- near the mouth of the remaining corridor.

The rapid warming at the onset of the Holocene did not produce any sudden changes in the Nordic Seas. The Polar Front moved to a slightly more northerly position along the eastern margin as the ice slowly melted. By 9,000 BP, however, large changes had begun to occur. The Polar Front, by this time, had migrated westward to a position near the Greenland Continental Margin northwest of Iceland, and for the first time since the Eemian Interglacial Period, there was a large presence of Atlantic water, which by then had become dominant in the Nordic Seas. By 7,000 years ago, the Polar Front had moved closer to the Greenland coast; sea ice was at its absolute minimum, confined mainly to the near coastal areas along Northeastern Greenland. The warm water of the Atlantic Current occupied much of the Nordic Seas, flowing like an enormous river through the eastern and central regions. The next 2,000 years was, of course, the height of Holocene warming.

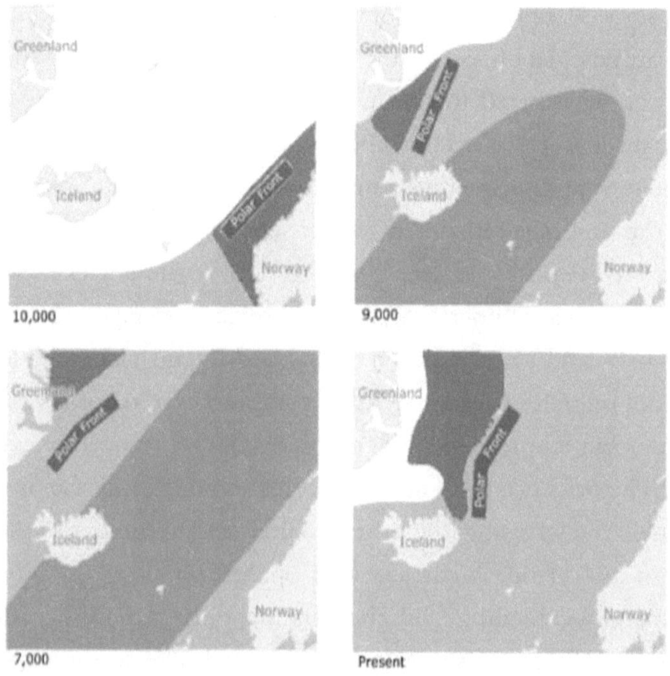

Presence of unmixed Atlantic water (medium shade), Arctic water (dark), sea ice (white), and approximate location of the Polar Front four times during its Holocene progression through the southern extent of the Nordic Seas. Numbers are thousands of years before the present (BP) (Koc, 2000).

By 5,000 BP, the water balance in the Nordic Seas began to change in favor of Arctic-derived water, and climate conditions began to turn back the other way. The Polar Front had rotated to a more north-to-south orientation, forming a boundary between a reduced but still significant presence of Atlantic water to the south and a growing influence of Arctic water that had already significantly increased in volume in the northwestern part of the seas.

The water balance in the last 5,000 years has significantly changed in favor of Arctic water. A couple of points discussed in Chapter 5 bear repeating. Winter sea ice reached its absolute greatest Holocene extent several hundred years ago during the Little Ice Age, but aside from that, it is still today at its greatest extent than at any time since the very early Holocene. And Atlantic water has decreased to the point where it no longer has any direct influence- not in the Nordic Seas, nor even in the far North Atlantic. Atlantic water- the Atlantic Current- today mixes with Norwegian Sea water in the northeast Atlantic- long before it enters the Nordic Seas. The Polar Front has returned to an east-to-west orientation; its lower portion now extends farther south and to the east of Iceland. The area along the Greenland seacoast where the front rested for much of the Climate Optimum is now dominated by Arctic water and by sea ice in winter. Perhaps the most recent important change is that in the last several decades, Arctic water, through the East Iceland Current, has become dominant east of Iceland, where it now flows directly- apparently permanently- into the central North Atlantic (Nabokov & Falina, 2006). Especially in a prolonged negative phase, southern and eastern margins in the Nordic Seas, in particular, become dominated by the influence of cold and fresh Arctic water, coming principally by way of the East Greenland and East Icelandic Currents. Today, the Polar Front has progressed a third or more of the way in its Holocene return trip across the Nordic Seas, and there is no telling how long the remainder of its trip will take. When the climate eventually reverts to glacial conditions, we can expect that this frontal system will have advanced again to its far southeastward position in the Nordic Seas (Koc, 2000).

By the end of the Climate Optimum, the period of warmth that characterized the Early Holocene had finally ended. Changes had begun several hundred years earlier and before, especially in areas farther north. Northern forests had reached their greatest extent during the early

Holocene; Scandinavian temperatures then were as much as 2° C (3.5° F) warmer than today, and tree lines, defining the farthest reaches of forested regions, were at some 300 meters (1,000 ft) higher altitude. Even as most areas were experiencing the warmer conditions of the Mid-Holocene Optimum, the warmth came to an end in Scandinavia as the climate became abruptly colder some 5,800 years ago.

A late Holocene return to a cooler climate

By 5,000 BP, a wider regional and global trend toward cooler and drier conditions had begun, signaling the onset of a new period of climate instability, declining temperatures, and glaciation that has continued, more or less, until the era of 20th-Century warming. The North Atlantic region experienced an increased influence of the East Greenland Current, advancement (migration) of the Polar Front, and an increase in precipitation and storm activity. Glaciers on Svalbard had retreated to at least today's limits by 9,000 BP, and during the Early Holocene, most glaciers in Norway and Svalbard were completely melted. But the Greenland ice sheet, having also receded during the early period of warmth, had started to expand by this time. And northern Iceland and Eurasian glaciers were growing again (McBean, 2004). At about the same time, climate conditions in tropical regions also began to deteriorate. The Sahara Desert in North Africa did not exist in its present form in the Early Holocene- the area was much greener than today with grassland vegetation. Instead of cooling, however, oceanic and especially continental temperatures in the tropics increased, and the climate in North Africa became much drier and dustier, with a rapid change to desert conditions around 5,700 BP. In the meantime, sea surface temperatures (SST) off the coast of West Africa began to decrease, an apparent response to a decline during the same period of the southern flow of deep Atlantic water, the upper portion of which- the UNADW- rises to the surface off of northwestern Africa, moderating the climate- and this is associated with the weakening climate conditions farther north. SST off the coast of West Africa had declined by about 5° C (9° F) by 5000

BP, as the climate began a more significant decline in the North Atlantic. The 400-year-long drought occurring at about 4,000 BP was extremely harsh across a wide area- from North Africa through the Middle East and into Southern Asia. Today, the climate has continued to dry, and the expectation is, with the warm and humid climate of the Early Holocene now long since gone, this is a trend that will likely continue. These areas, in particular, and other landmasses, including large areas of Australia, can be expected to eventually return to the more harsh arid environments that existed there during the last glacial period (deMenocal, 2000).

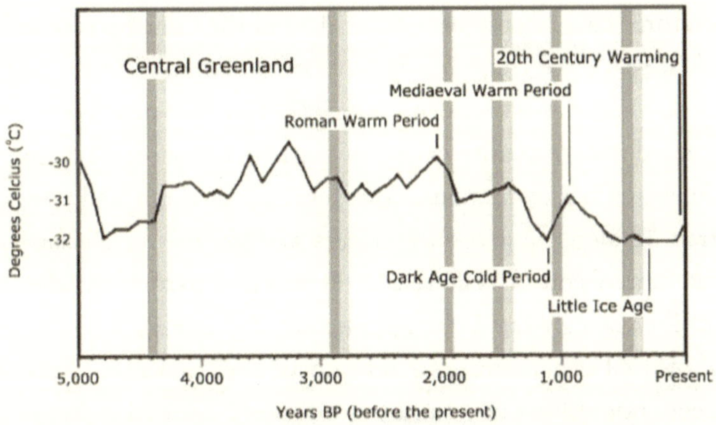

Increased frequency of glacial events in the Late Holocene: Periods of glacial advancement in northern Iceland are indicated by the darker gray vertical bars; sea ice (ice-rafting events) in the North Atlantic-light gray bars. Note the stronger trend toward cooling in the last one to two thousand years. Climate events, some of which are specific to Europe, are included to provide some general reference points. This graph shows changes in temperature of ½ degree or more. The data is sourced from the NOAA and GISP2 Ice-Core (Alley, 2004).

Summer temperatures are important in determining long-term trends, and in the Arctic, temperatures have generally declined for the last 3,500 years. In the Canadian Arctic, a decline is evident for the entire period of the Late Holocene, and tree-ring chronologies in northern Siberia show a significant and progressive decrease in growth, indicating a declining trend in temperatures for the last 6,000 years. In the Baffin Bay area of Northeastern Canada, the period of Early Holocene warmth occurred later; temperatures there warmed up by about 8,000 BP and became even a little warmer by 6,000 years ago. Since then, however, the climate there has cooled- cooled markedly by 3,000 BP, and by about the same time, a renewed period of cooling and glaciation had begun in Alaska. Particularly severe conditions prevailed in northern Scandinavia as the climate there cooled from about 2,600 to 2,000 BP. This was a time when temperatures were both generally cold and highly variable and when glaciers were at their maximum Holocene extent (McBean, 2004).

Late Holocene cooling events have occurred in the North Atlantic at fairly regular intervals of approximately 500 to 1,000 years. Cooling temperatures have been especially evident in the Canadian Arctic in the last 2,000 years; the trend has been more pronounced in the last 1,000 years and even more so in the last 500. The climate, in general, has deteriorated over the entire period of the Late Holocene; the decline has been anything but continuous, however. Expected ups and downs have occurred all along, but cooling events and periods of glaciation have become more frequent. The last 1,300 years, in particular, have been a time of increasingly rapid change; similar, it is interesting to note, to a period of oscillating climate found to have occurred at the onset of the Holocene. For at least the last several hundred years, the climate in the Arctic and the North Atlantic region has become much more variable and capable of more extreme changes- increasingly oscillating between warm and cold conditions. A number of glacial advances in northern Iceland have occurred in the Late Holocene, and as time goes on, they are occurring with increasing frequency. Several of these advances correlate with ice-rafting events in the North Atlantic, which are also found to be occurring with increasing frequency. The Little Ice Age was, of course, characterized by the most severe climate conditions of the Holocene, most severe in the Arctic, but its influence was global. Preceded by a century of ice-rafting, glaciation occurred worldwide, reaching its greatest extent since the Younger Dryas cooling event. The coldest temperatures of the Little Ice Age, as previously stated, occurred in Northeastern Canada only a couple of hundred years ago. Thus, today, after a century of warming, we are still not very far removed from the coldest conditions in at least 11,000 years (McBean, 2004).

North-to-south change in Insolation

The climate of the entire Holocene can be understood within the context of latitudinal changes of warmth and direct sunlight, or Insolation, at the Earth's surface; the direct result, of course, of the Earth's slowly changing orbital conditions. A particularly close correlation is evident- like hand and glove- with Insolation and Precessional Cycle. At the beginning of the Holocene, when the Earth's northern polar axis was more directly tilted toward the Sun, and this coincided with Summer Solstice occurring at the point of perihelion, the Sun's energy was more concentrated in the high northern latitudes, about 465 w/m2 (watts per square meter) at 65° North- the latitude of Iceland- and so provided the warmth needed to raise temperatures above glacial levels. Since then, the focus of this energy has very slowly, very gradually, moved southward so that today, Insolation is stronger in the tropical regions and in the Southern Hemisphere, and the Sun's energy in the far north has now declined to approximately 422 w/m^2. The progression of SST over the period shows this as well. Both the North Atlantic and the Tropics warmed toward the Mid-Holocene. However, the overall progression in the North Atlantic since the initial warming has been toward a progressively cooler climate.

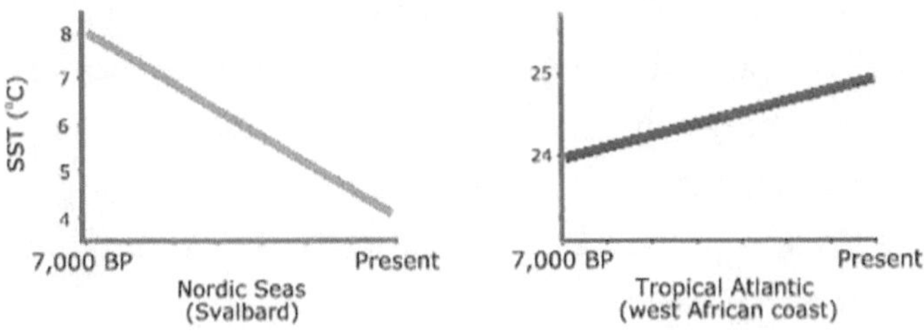

The north-to-south Holocene change in Insolation since the AIM 7,000 years ago. Small temperature changes in the oceans applied globally make all the difference in the climate. Data is derived from fossilized remains of ancient surface water plankton recovered from deep-sea cores (Rimbu, 2004).

In contrast, at the same time, at tropical latitudes and in the Southern Hemisphere, temperatures have warmed. Orbital changes in Insolation are especially significant at high latitudes, and 65° N is the latitude that appears most sensitive to changes in summer Insolation. This latitude transects the region of the Subpolar Seas between Greenland and Norway; it is at the very center of what is perhaps the most critical link of the Thermohaline Circulation- of ocean circulation. Along this latitude is the point at which ice sheet formation began and ended during the last glacial period. So, with respect to climate, it may be the most critical region on the planet- the place where ice ages begin and end (Adams, 1999; McBean, 2004). The effects of the current warming- the heat waves, the droughts, the dust storms that sometimes accompany them, increased storm activity, and flooding- are all increasingly evident. But to put this warming into perspective, it must be measured against at least 5,000 years of a generally cooling climate- and this is especially true in northern regions. Although interrupted early on by the cold Younger Dryas, the initial warming of the current interglacial period really began about 15,000 years ago, and, as indicated in Chapter 4, by the time the Eemian period had progressed to this point, the climate was ready to return to glacial conditions. Now, this in itself does not mean that an ice age is about to begin, but we have to take it into account. The changes observed in ocean circulation in the North Atlantic and in the Nordic Seas in the last several decades were also in play at the end of the Eemian. Another factor that adds to this concern regards the rapidity that is evident in large climate change. There will be more about this in a later chapter, but an analysis of change events during the last glacial period, as well as transitions that have occurred before between climate states, suggests that an unprepared population would be severely strained to adjust to the rapidly evolving new conditions. The progression and recent changes in the water balance in the Nordic Seas are another indication of where we are in the glacial cycle. To put this in an even broader context, the Earth's orbital conditions and the resulting changes in Insolation suggest that it is now quite late in the Holocene, and these conditions do not favor an extended period of warming in the Northern Hemisphere. These are significant and long-term climate signals, and they provide a basis on which we can attempt a more complete understanding of present conditions and to better anticipate future changes.

PART THREE

UNPRECEDENTED CHANGE

7

DRAMATIC EVENTS IN THE NORTHERN SEAS

Weakening ocean processes-
Atlantic water loses its great northern arm

Significant changes have occurred in ocean circulation, and there may be no single climate event in modern times that has greater long-term implications than the decline of the convection process in the Nordic Seas. The weakening has occurred throughout the region and through the entire water column, but most notably in the Greenland Sea, where its total strength today is less than 20% of what it was in the early 70s (Cottet-Puinel, 2004). Among the most important highlights is a near complete cessation of deep-water convection there in 1994.

Today, convection in the Norwegian and Iceland Seas is limited to intermediate levels; only in the Greenland Sea does the mixing occur at depth. Following the formation of the temperature and salinity barrier there, produced by the Arctic surge of the early 90s, it occurs today only in the very center, and sometimes at the periphery, of this once massive convection region. Consequently, there is today a much-reduced contribution to the deep southbound currents that are critical to maintaining the relatively warm climate conditions that have prevailed during the Holocene Period (Karstensen, 2005).

The importance of the Greenland Sea convection process in maintaining our place in the glacial/interglacial cycle of change cannot be overstated. Atlantic water's huge dominance and warmth in the Nordic

Seas during the early Holocene was possible solely due to its single influence. That there has been any warming at all at this point in time is mainly due to the revving up of its vast overturning of Atlantic water at the dawn of this period of warming.

With the rather sudden downturn in the strength of that great climate machine in the Greenland Sea that ended the Early Holocene some 5,000 years ago, the process was picked up in the second most important area for the overturning process- the Labrador Sea. Convection was strong in both seas at the dawn of the Holocene, as the climate first emerged from ice age conditions. However, the Labrador Sea had fallen silent early on, importantly due to widespread glacial melting and a resulting freshwater cap that became established there. An inherent aspect of the global ocean circulation system involves its reestablishment in the event of a disruption in an ocean current. It can be thought of as a kind of built-in fail-safe device that nature employs when needed. So when a disruption occurs, like what we see now, especially in the Greenland and Labrador Seas, it will always re-connect somewhere else.

There are many climate events in the Earth's recent history that can be described as unprecedented. In this analysis, we primarily single out two events that have occurred in just the last few decades: two significant climate shifts that define this era of change. And these are, as we have already identified, the mid-nineties event and the next one beginning in 2010. The climate record reveals that the circulation of warm tropical-derived water north of Iceland and into the Greenland Sea has been the great northern arm of Atlantic water during this period of warming. Today, it is a mere shadow of its once mighty reach into the Nordic Seas and beyond.

The primary entry points for southbound water into the Atlantic-Denmark Strait and the Faroe Channels- have weakened correspondingly. Both entry points occur along a subterranean ridge system- the Greenland-Scotland Ridge- that defines the southern boundary of the Nordic Seas. Compared to its strength during the 1960s, the deep-water portion of the current through the channels has significantly changed- found by the mid-90s to have declined by some 50% (Østerhus, 1999), which is consistent with the weakening of the Deep Atlantic Current, described in the following pages. Changes in the currents at these two overflow points reflect a decreasing contrast in the relative densities of Nordic Sea water and the North Atlantic. It occurs along the entire span of the ridge system and is a direct result of ongoing warming and freshening in the upper and deep water layers. This is important because the relative differences in temperature and salinity are critical to ocean circulation. This weakening is due in no small part to the loss of the cold salty water contribution from the Greenland Sea, where the deep water, in particular, is the coldest in the Arctic (Hansen, 2001; Cottet-Puinel, 2004). Colder even than in-flowing Arctic water from the north, which seems contradictory, but deeper usually means colder, and at the bottom of the Greenland Sea is, as stated before, the deepest ocean basin in the Northern Hemisphere.

As the water from the Nordic Seas flows across the ridge and into the Atlantic, recent studies show it further weakened by some continued mixing- now with the cold freshwater that has become widespread in the Atlantic basin. Separating into individual currents and continuing downstream, the deep water- the Deep Current- from the Faroe Channels, otherwise, maintains its chemical integrity. Eventually making their way into the Western Subpolar Gyre and the Labrador Sea area, they combine with water from the convection process there.

Besides the Labrador Sea, the Irminger Sea and a smaller region that has formed more recently off the southern tip of Greenland also contribute to the convection process at intermediate depths. In recent times, deep convection in the North Atlantic has occurred only in the Labrador Sea, which, since the early 90s, has been limited almost solely to an area along the Canadian continental margin. And in-flowing Arctic water, along with melt-water from Greenland, has produced a barrier there, at the bottom of the intermediate layer, similar to that in the Greenland Sea, and at the surface, a freshwater cap- like what happened there in the early Holocene.

As you might recall from the description of this process in Chapter 2, all this water forms two great underwater rivers -an upper portion of intermediate water and a lower one comprising the Deep Water Current. Flowing southward from there, they are collectively referred to as the

Deep Western Boundary Current. With a weakened contribution from the convection process, especially at depth, and an increasing influence from Arctic-derived water, this great current begins its long trek toward the tropics and beyond. It never-the-less continues to maintain its layered integrity- like that found throughout the oceans. Being somewhat sandwiched between the northward-moving Gulf Stream at the surface and another northbound current of Antarctic origin near the bottom, this great underwater river lies between 2,000 and 5,000 meters in depth (6,500- 16,500 ft). The upper part of it, down to about 3,000 m (10,000 ft), consists of water from the Subpolar North Atlantic region and has maintained approximately the same volume since the current was first measured in 1957. It continues into the tropical Atlantic, where it rises to the surface.

The decline of the Atlantic link
in global ocean circulation

The lower portion of this current, below 3,000 m, comprises the Deep Water Current; it eventually surfaces in an area near Antarctica, where it exerts a moderating influence on the colder Antarctic climate. Moreover, it is this deep portion that is the most important, for it alone completes the Atlantic link, connecting to the global system of surface and underwater currents, a system of circulation that provides the single means for the distribution of heat, salinity, and nutrients throughout the world's oceans. Its volume today is found to have decreased compared to 1957, with the most significant decline occurring during the mid-90s. This part of the current has its sole origin in the Nordic Seas, and its weakening is the direct result of the declining convection process there. This is a reason for concern, for the decline of the Deep Current and the corresponding weakening of the Atlantic Current at the surface are associated in the climate record with the onset of periods of climate cooling. Should this trend continue, it could lead to a significant weakening or even a disruption in the ocean's circulation system, initiating expectedly large climate changes worldwide (Bryden, 2005).

The measurements and the data collected, and their interpretation during this multi-decade study period, are consistent with climate changes observed in the North Atlantic and Subpolar Seas. In 2004, more extensive and continuous measurements began, which revealed a seasonal variability in the strength of the Deep Current. Significant voids occur, particularly at depth, and most notably in November of 2004 when it stopped completely for ten days. Its sharp decline, evident in the '98 measurements, is certainly consistent with the weakening of important aspects of the climate system in the North Atlantic region at the time. The climate farther north continues its broad-based decline. Neither has there been a recovery in a loss of a lower 400 m (1,300 ft) portion of the Deep Current; found to be missing in the '98 measurements, it has not been regained (Bryden, 2005).

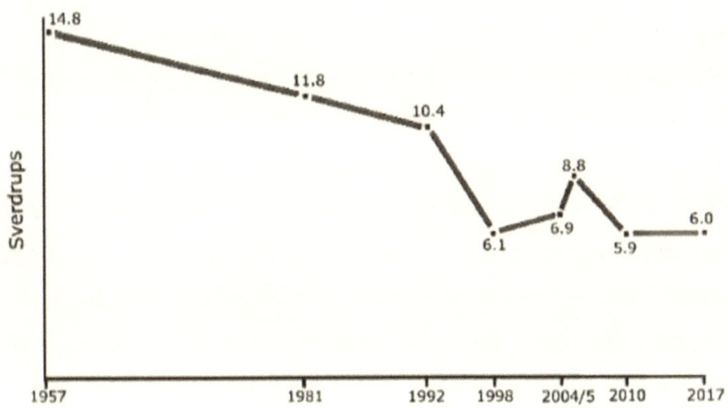

Measured continuously since 2004, this graph highlights the volume changes (approximate values) in the Deep Current (LNADW). Expressed in Sverdrups (the unit used to measure ocean currents): 1 Sverdrup= 106 cubic meters per second, the same rate as the entire volume of freshwater flowing from the world's rivers into the oceans (Bryden, 2005; McCarthy, 2012; Smead, 2018).

The changes in the North Atlantic region result from an interaction of atmospheric and oceanic influences; a feedback mechanism has developed as each changing aspect influences each of the others. The last several decades reflect an increasing oscillation of climate events and a changed relationship between the NAO and important climate aspects. Sea levels, for example, determined by temperature and wind strength, have not, since 1996, adjusted to their expected values in the North Atlantic (Esselborn, 1999). The strength of circulation in the Subpolar Gyre, of which sea level is an important indicator, continues to weaken. The sea-to-air heat transfer rate, which characteristically weakens when the NAO changes to a negative phase, is expected to strengthen again as the climate returns to a positive one. But it instead very uncharacteristically began to weaken in the early 1990s, at the height of a very strong positive phase. And with the intense negative shift in '96, heat transfer and convection are in steep decline. Of particular concern is that the NAO has changed phase several times since then, and the decline of these important climate aspects now continues regardless of which climate phase is dominant (Satellites et al., 2004; Häkkinen & Rhines, 2004).

With the continued southward flow of Arctic water, the temperature, chemistry, and structure of the water column in the Greenland Sea is becoming indistinguishable from that of the Arctic Ocean. Density contrasts necessary for strong ocean circulation are in decline in the entire region due to resulting changes in temperature and salinity. And, as time goes on, as a consequence of these events, the reality posed by them is more and more apparent. One wonders, in fact, with the Sun's energy at the low point in the Insolation cycle in the Northern Hemisphere, where the energy would come from to produce a hoped-for reversal of these changes.

With the expanding influence of Arctic water in the region, the volume of Atlantic water is decreasing. And even though temperatures in the in-flowing Atlantic water are still strong, we should expect the continuing weakening of the current to eventually result in less heat delivered to the northern regions- a scenario that, after the events of 2010, has become a reality. Temperatures are always changing, but nowadays, they seem to be changing a lot. And those changes do not necessarily reflect a trend toward the 'Global Warming' that we hear so much about. There are many instances of unprecedented change.

Where there's warming and where there's not

In the Arctic, by 1850, temperatures began to increase as the ice began to retreat. Two periods of significant warming occurred during the 20th Century. The first one peaked in the 1930s, giving way by mid-century to a cooling trend; then, in about 1970, a new warming began that increased into the 1980s and has continued to the present- although there are signs that may be changing. Globally, temperatures since 1900 have increased about 1° C (almost 2° F) on average. This may seem like a small change, but small changes can produce large climate effects. The warmest times of this interglacial period, the Mid-Holocene Climate Optimum, was only about 2° C (3.5° F) warmer than today. The warming, however, is not evenly distributed; much of it has occurred in the tropical regions and, more recently, in the high latitudes. Subsequently, overall, the change in Arctic temperature is at least as much as the global average, with some studies showing near-surface air temperatures at land-based meteorological stations to have increased almost 2° C. In other areas, it is even more- there is much variation within the region. Generally, there are many shallow coastal areas where sea-surface temperatures (SST) have increased for several hundred years, but the recent warming has occurred primarily in continental land areas.

Most of the North Atlantic and Nordic Seas began to warm up by 1970 or '80, but before that, temperatures there were in decline. And over the Twentieth Century as a whole- until the more recent appearance of the Blob, which will be covered in the next chapter- there has been no significant trend toward warming or cooling. In the western and northern parts of the Nordic Seas, SSTs are not warmer today than in the early 1900s when the warming began. In the eastern Nordic Seas and the northeastern Atlantic, SSTs were in decline until warmer, saltier water began flowing into these areas through the Atlantic Current in 1996. This cannot be said to necessarily apply in a broader geographical sense because the temperature increases that have occurred in the Arctic in the last 50 or 60 years may be without precedent since the early Holocene, but in the Nordic Seas, it appears that this latest period of warming is so far found not to be remarkable as compared to the warming during the early 1900s, or to those of the earlier Medieval and Roman warm periods- both of which were, in fact, warmer than today. More recent measurements, since the changes in 2010, however, as we will later see, show a new trend towards cooling- first in the North Atlantic and then extending into the Nordic Seas.

We all accept that the Earth's climate is generally warming, but a closer examination of what this means suggests that the warming is not as clearly defined as we might have thought. The results of any temperature analysis will depend upon the time period and geographical area selected, as well as the type of data included. If eastern Antarctica was the only area sampled, then you might not know that temperatures were warming up at all. There are a number of meteorological stations on the continent, and some of them register the absolute coldest temperatures on Earth; most of these stations show that temperatures there have actually been getting a little cooler, especially in summer, over the past few decades. It is along the coasts of Antarctica's Western Peninsula, where some pretty big pieces of the great ice shelves have broken away that there is melting and rising temperatures. But it may be surprising to know that the cause is geothermal, in other words, volcanic- instead of a general warming. Ice growth on the East Antarctic continent itself is actually found to be increasing. In the Northern Hemisphere, temperatures are increasing, but not more, if as much, as the warming that occurred in the early part of the Twentieth Century. Most of the warming has been in coastal areas where the rapid melting of Greenland's coastal glaciers has been underway (Taylor, 2006; Box, 2009)- later chapters will reveal new changes that suggest the warming may in fact be coming to an end.

An increase in the global average temperature does not necessarily mean hotter summers; it does mean that daytime temperatures are more often in the upper part of the average temperature range. In other words, it means more hot days, but not necessarily hotter ones.

Our examination might also bring us to question the impact of the increasing amount of atmospheric moisture on temperatures. We know there is more moisture in the atmosphere these days; this is what the strengthening Hydrological Cycle is all about. We need to keep in mind that water vapor is the most important greenhouse gas, and its increased presence in the atmosphere is the reason for the increase in temperatures within the daytime temperature range- the DTR. An effect is particularly apparent at high latitudes in the Northern Hemisphere, where significant increases have occurred in atmospheric moisture.

CLIMATE SHIFT #2

Invasion of the Cold Blob!

A new dominance of cold fresh Arctic water in the Iceland Basin region of the northeast Atlantic- an area called the Eastern Subpolar Gyre- reveals a critical climate signal. It is of particular importance because the climate record- the record of the past- indicates that similar events there in the past have heralded an imminent return to glacial conditions. Referred to as a 'warming hole,' official sources, including NASA, seem to want to refrain from using the word cooling in their public pronouncements- but a cooling is exactly what it is. Accompanied by the most significant decline to date in the Atlantic Current, these two events together comprise a second and much more visible shift in climate conditions. The length and breadth of this cooling include essentially the entire region of the Subpolar North Atlantic, consisting of an eastern and western circulating gyre. Its influence extends south, even into the subtropical zone- to about 40° N Latitude, and a little beyond- that also consists of east and west circulating gyres.

It is associated with a significant decline in atmospheric temperatures, strong enough to have diverted the Northern Jet Stream to the south and east around it. A record low in the NAO in 2010- an important signal- preceded these changes. There was an even more marked decrease in sea surface temperatures. At its central point, south of Iceland and Greenland, surface temperatures cooled as much as 2° C (3.5° F). It is more than a mere surface phenomenon; its depth extends down more than 1,000 meters (3,300 feet) and includes essentially all of the upper ocean layer. At its beginning, scientists gave it the name 'Cold Blob,' which seems appropriate because there are some ominous implications to its persistent presence.

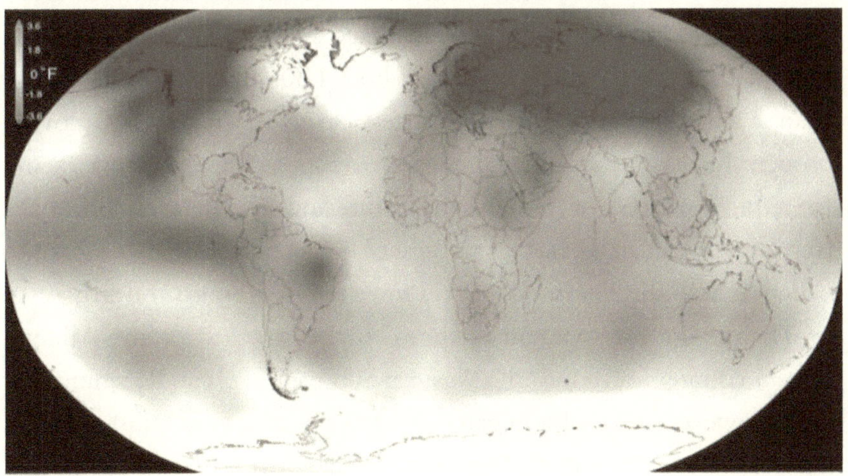

The white in this NASA image represents areas that are cooling, and shaded areas warming- of particular significance is the subpolar region of the North Atlantic.

There is an old science-fiction thriller movie entitled 'The Blob;' this could turn out to be the monster that the name Blob implies. More than a cooling, this is also a colossal freshwater event. It is, in a word, unprecedented, and it is the most recent in a series of such unprecedented events that have characterized the climate in modern times. Unprecedented, that is, in the historical record, such events are known to have occurred before. They are evident in the climate record- in the world's sediment, rocks, and glaciers.

The initiating event that brought about the appearance of the Cold Blob reached a critical threshold in the winter of 2009-2010, with a rather sudden and dramatic thirty percent weakening in the North Atlantic Current. This is an event that has come to be understood as a response to the previous decades of declining strength in the south-bound Deep Current, as well as an increasing recirculation of Atlantic water back southward into the Subtropical Gyre (Bryden, 2020). Thus adding to the warming there and increasing the intensity of storms forming off the west African coast.

Of particular interest in this respect, along with the other things going on in the region, is a heaving of the colder lower ocean, resulting in a rise up into the surface layer of the Thermocline. The Thermocline is a boundary area between the relatively mixed surface layer and the colder deep ocean below it, which resides at about 1,000 m depth in the North Atlantic. Its rising makes it more accessible to those currents moving through the ocean above it, and ocean currents will tend to move in and spread out into areas of similar temperature and salinity levels. So what's happening is that an increasing amount of the recirculating Atlantic water is slipping into the Thermocline and traveling back south at depth.

There were no apparent changes to the strength of the Atlantic Current coming north into the Western Subtropical Gyre. Reaching that point, however, some of it began to spread out westward, piling up, you might say- along the Atlantic Seaboard of the North American Continent. Some more of it- an increasing amount- turned east and southward, back into the Subtropical Gyre. That still left a lot of warm water going north, but these conditions have persisted, and as time goes by, there is less and less Atlantic water and its warmth going on north.

In a little over a year after its initial weakening, the Atlantic Current had recovered its strength but has since then been in a steady state of decline- a decline expected to accelerate. At any rate, the North Atlantic Current is the single source of heat for the waters and especially coastal land areas of the North Atlantic and beyond. As a result of the loss of all this warm Atlantic water, there began an equally dramatic ocean temperature decrease, especially in the Subpolar region- and a wide-spread expansion of colder water of Arctic origin.

The first signs of our cold Arctic-derived Blob began on the western side of the Atlantic, most prominently in the Western Subtropical Gyre, directly in the way of the northbound Atlantic Current, as evidenced by ocean temperature measurements. Then, by 2012, a significant accumulation had begun to form in the eastern margins of the Labrador

Sea. There is more than a small amount of uncertainty about the origin of all this cold and fresh Arctic water because there was just so much of it. Some of it was newly arrived by way of the East-Greenland Current and perhaps through that relatively new access point for Arctic water east of Iceland. Much of its expansion is thought to be due simply to an accumulation within the Subpolar region itself, due in part to a couple of unusually severe winter seasons that occurred during the time of its formation in the Eastern Subpolar Gyre.

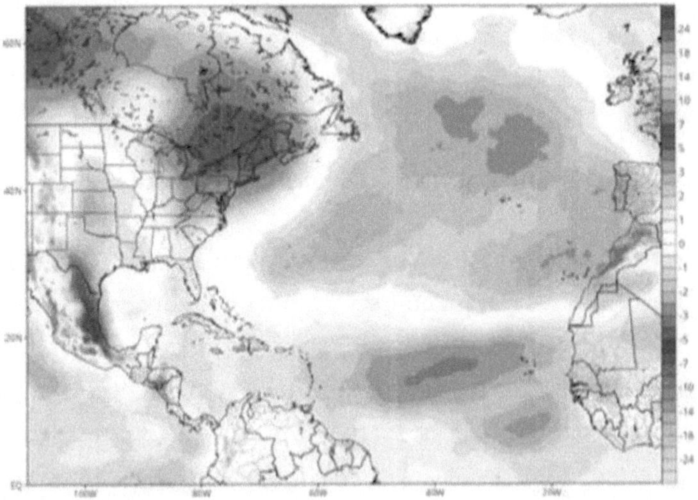

The Blob's new dominance is evident in this NASA image from 2016. The darkest areas toward the north represent the coldest conditions- and farther south, accumulation of warmer Atlantic Current.

From the Labrador Sea area, its progression could be followed south and then eastward, all the while maintaining its own temperature and salinity levels. Increasing in size as it went, it entered the southern aspects of the Eastern Subpolar Gyre, slowly displacing a newly weakened Atlantic current to the south and farther eastward until finally becoming established in the northeast Atlantic's Iceland Basin, where it has remained. In scientific terms, it is a great temperature and salinity anomaly; it is a mass of water, by one estimate, of approximately 6,600 cubic kilometers in size- or nearly 1,600 cubic miles (N. Holiday, 2020). And to give an idea of how much that is, it is enough water to cover the rooftops of most single-story houses and buildings in an area the size of the eastern coastal region of the United States, from the Atlantic to the Appalachian Mountains- about 5 m (17 feet) deep. It is an event, the significance of which far exceeds any comparable one on record, including the 'Great Salinity Anomaly,' that moved through the region in the late 1960s and 70s.

We can expect changes over time, but make no mistake- that Blob is still there. In today's climate, when Arctic water gains some ground, so to speak, it tends to keep it. We should not count on our Blob going anywhere anytime soon- it could turn out to be a permanent fixture. A more significant event than its 1960s predecessor, it is associated with the most dramatic change on record in the North Atlantic Current.

Warm water takes a back seat

This Blob has been in its present configuration, centered in the Iceland Basin, since 2015. We cannot now say whether this is a permanent change, but we need to ask ourselves what it means for us as we search for a way forward. How does this play out with the Atlantic/Arctic water balance? When you think about it, at least for now, this amounts to a rapid expansion of Arctic water and replacement of Atlantic water in the Subpolar North Atlantic.

With the arrival of the Blob, the northward-moving Atlantic Current has lost its dominant position. With the Blob occupying the basin, Atlantic water is confined to a much narrower pathway along the European continental margin; some of it has even found a narrower path north around the Blob's western perimeter. This new influence of Arctic water extends from the far north-east Atlantic to the coast of Newfoundland. Essentially, a new re-organization of water masses has taken place. The Atlantic Current has become relegated to a secondary role for the first time since the last ice age. And Arctic water- a good portion of it unmixed- is the new dominant player, so to speak, in the North Atlantic.

Like in the Nordic Seas, there is an atmospheric frontal system called the Sub-Arctic Front, which forms the dividing line between Atlantic water- in this case, the Atlantic Current- and the Blob, or water of Arctic origin. It will be found right alongside the current- situated between it and the Blob. As the current has become displaced to the south and east of its former location, it will be interesting to see what this might mean for the farther north Polar Front in the Nordic Seas and its important slow eastward Holocene journey toward a glacial climate.

Thus, we must also ask ourselves the broader implications for the Earth's climate. What are the implications for some eight billion human beings that live here? What, in the worst-case scenario, might we expect from this particular event?

To put it in just a few words, nobody knows precisely how things will unfold. It is possible that this Blob is the beginning of an ice age- or it may not. We might see a few of these come and go as time passes. It is, however, the kind of event found in the climate record to have in the past preceded by only a short time- as little as ten years- the onset of severe changes towards glacial conditions.

A large area of cold and fresh circulating water called the Beaufort Gyre, accumulates at the surface in the western Arctic. Typically, it has reversed direction every six years or so and, in the process, released some of its water into the surrounding Arctic. Lately, though, it has gotten off its cycle; it has been at least a dozen years since it has changed directions. The concern is that when it does, a large volume of accumulated cold fresh water will come out of it.

As mentioned in the introduction, the Arctic Ocean, surrounded mainly by land, acts like a big bathtub. As more and more water is added- mainly due to the acceleration of the Hydrological Cycle, which we will spend some time on in the next chapter- it will overflow. And you can bet that a lot of this water, whether it is from the Beaufort Gyre or a generally increasing Arctic precipitation and melting, will find its way through its only outlet- south and into the North Atlantic.

By the time our Blob had reached its greatest extent and coldest surface temperatures, global warming- in that part of the world anyway- had become harder to find. And as time goes on, that Arctic water access point east of Iceland, around the Polar Front's southern flank, will expand.

Between it and an ice age world there is now nothing but the open sea. When it is no longer just a holiday, when the warm Holocene is finally gone for good, we had better be prepared.

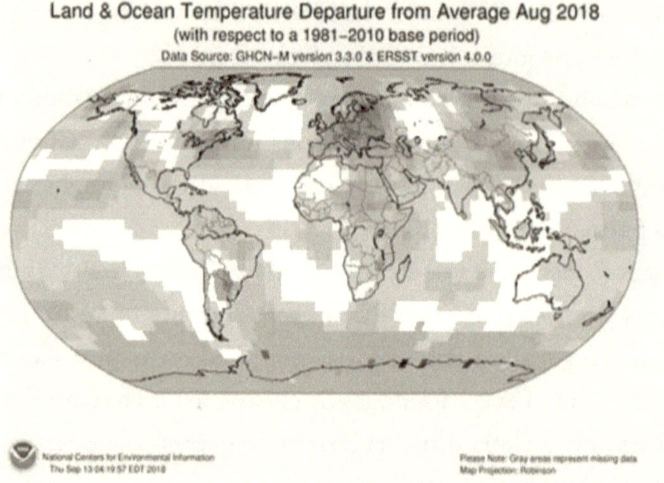

NOAA image of the height of summer warming in 2018-an excellent example of what might be called a global warming holiday! White areas indicate cooling.

It is also interesting to note that the cooling effect of this latest Arctic cold trend extends onto the northeastern Canadian mainland, including the many islands that make up the archipelago in the far northeast and south from there to Newfoundland. All of this coincides with this new expansion of the influence of cold, fresh Arctic water. It may be no coincidence that this same area of northeastern Canada was also the coldest at the time of the Little Ice Age and the last place to warm up as the cold climate that characterized that period came to an end.

One problem we run into in our approach to the Cold Blob is accounting for its massive size without more of a direct contribution of additional water from the Arctic. One has to wonder, even though no direct evidence for it is known, whether or not there is a connection between the cooling in the northeast Atlantic and the very similar cooling, at the very same time, in the far northeast of Canada. One has to wonder, the volume of the Blob being what it is, whether or not some Arctic water made its way, unnoticed, through the archipelago, or through its new access point to the Atlantic east of Iceland to join in its formation.

As of 2014, a new array of ocean buoys have been set up in the far North Atlantic, from the southern tip of Greenland to Ireland- the entire span of the Greenland-Scotland Ridge- and across the Labrador Sea, from western Greenland to Canada (Lozier, 2019). So, in the event of some future blob's appearance, we will know what contribution a direct flow of Arctic water from the north has made.

That such a cooling and freshening is occurring now, at this particular time and place, should be of special importance to us. There are some important consequences of which we need to take notice. The Atlantic Current has been running hard and strong since the beginning of the Holocene. Its great strength reached all the way up into the Nordic Seas and beyond for much of this time. And now, since that first great shift in modern times in 1996, as mentioned before, it has lost its great northern arm. And now comes this Cold Blob that has positioned itself directly in its path. In the absence of our Atlantic Current in this region, there would be no Holocene warming; we would be left with only the cold to talk about. It is easy to see that the weakening of the current preceded the development of the Arctic Blob in the years following 2010. It is also apparent, however, that there was some accumulation of Arctic water, even before that, directly in the way of the current. It's rather like the question of which came first between the chicken and the egg. And now the current finds itself taking something of a backseat to the Blob's expanding influence.

The Atlantic Current moves to re-establish itself

This is more than a little concerning- it is a huge deal! Nevertheless, these are the facts. Already weakened from decades of an increasing presence of Arctic water, the Atlantic Current finds itself rather suddenly in a further weakened state, accumulating along the North American coast and turning back south to join an increasingly recirculating Subtropical

Gyre. And this is not new. It has been apparent since at least 2005. What is new is that there is more and more of it. Of particular interest is that its departure from the main current is in the same general area in the Atlantic basin where the overturning circulation has re-established itself at the end of previous interglacial periods.

It is increasingly evident that this recirculation is a part of the problem. With each change in the region, there is a corresponding increase in Atlantic water turning back south, either at the surface or at depth through the Thermocline, and then returning into the Subtropical Gyre (Bryden, 2005 & 2020; McCarthy, 2012). What is apparent, however, is that the new decline in the Atlantic Current north of about 40° N, and a resulting reduction in the heat being carried from there on north, is due in no small part to this recirculation. The conclusion to be drawn from this is that the dreaded re-establishment of overturning circulation in the Atlantic has already begun and is increasing- and, in fact, has been underway for some time.

9

THE BIG PICTURE

What is wrong with the official story?

Because carbon dioxide holds a central place in any discussion or debate on the climate today, it is necessary to examine the evidence to determine, as best we can, what role it actually plays in the broad scheme of change.

Carbon is an element that is in abundance in the Earth's crust-its surface. It is everywhere around us: the rocks, the soil, the trees-essentially everything we see in the natural world, and much of what is in the homes in which we live is composed of carbon. We ourselves, besides being mostly water, consist of just four carbon compounds. In the atmosphere, in its gaseous form as carbon dioxide, however, there is hardly any of it, and that is the first problem with the generally accepted carbon-caused explanation for climate change. Its atmospheric content is often expressed in parts per million (ppm)- the measure used in science for a small quantity of any gaseous element. 300 ppm is the accepted atmospheric level for carbon dioxide at the start of the industrial age, in about 1850, taken as the beginning point for a greater human input into increasing amounts of carbon in the air.

Today, that level is as high as 420 ppm- about 40% more than its pre-industrial level. And that looks like a significant increase, but we need to break it down and see what it means. To convert ppm to a percentage, you divide by 10,000, so as a percentage of the total atmosphere, that amounts to just 0.04%, or expressed as a fraction, 4/100th of one percent- a tiny amount. It follows then that its total increase would be only 0.01% of that, or 1/100th of one percent- an even tinier amount. And we're not done yet because most of the increase is considered to be due to natural causes- estimated as high as 95%. We do not know the figure exactly, so let's take half of that as an estimate of the human contribution. So now we have 0.01 multiplied by 0.50 (50%)= 0.005, or 1/200th of one percent- a ridiculously small amount. If we instead use the amount of increase, the 0.01%, in our calculations, it would still not be enough to make a difference. Not even the total amount of carbon's atmospheric content of 0.04% would be enough to cause an appreciable change in temperature.

If we lived on Venus or Mars, where the planetary atmospheres are primarily carbon dioxide, then it would be different. But we're not; we're here on Earth with a mostly nitrogen/oxygen atmosphere, a little argon and water, and only trace amounts of carbon dioxide- and even lesser quantities of some other elements, like methane and nitrous oxide.

Carbon dioxide truly is a greenhouse gas, but water, as water vapor, is also a greenhouse gas- and a much more powerful one. Water is considerably more reactive with the other elements, so it has a greater role to play in chemical interactions, and it is present in the atmosphere in much greater quantities. At a little over 2% on average and as much as 4% of the total atmosphere, there is often 70 to 80 times as much water in the air as carbon. A contest between the two would be a pretty uneven match. Such a match-up would likely result in a very different ending from, say, that of the ancient Hebrew story of young David slaying the giant Goliath. In this case, comparatively speaking, if our young carbon were, let's say, the size of a strong 6 ft tall warrior, our giant water would be of such a size as to easily reach over a modern 40-story building.

The climate record clearly shows that it is temperature, not carbon, that leads the way in times of change. That temperature is the leader is especially evident at large change events when the climate makes a switch between glacial and interglacial conditions. The deepest and longest (oldest) record of such change is from the Antarctic Vostok Station Ice-core. It reveals a follow-the-leader relationship between temperature and carbon.

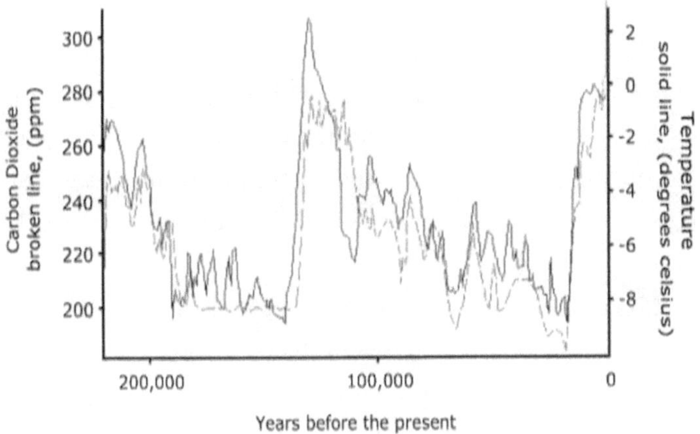

Changes in temperature and atmospheric carbon over the period of the last three glacial/interglacial cycles. Note that at each point when an interglacial warm period ended and the climate returned to glacial conditions, temperatures declined first - thousands of years before the level of carbon did. The data source for constructing this graph is the NOAA, collected from the Vostok Ice Core in eastern Antarctica. Each point on the graph is an average of 1,000 years of data. The zero point on the temperature scale, which is the basis for comparison, represents a 1950 measurement of −55.5° C (-67.9° F). The end of the carbon data from this core is about 1850; its rapid increase has occurred in the last 50 years (Petit, 1999).

As the climate transitions from one state to the other, the temperature is first to change. Their correspondence is closest at times of warming, and when the climate cools down, carbon still tends to follow along behind- usually several thousand years later. Regarding major climate events, the beginning of the Holocene represents the closest correlation between the two on record. A search for instances where carbon dioxide levels in the air increased ahead of temperature change reveals that in no instance, regardless of whether the climate was warming up or cooling down, did carbon lead the way.

The next aspect to call into question with the official version is its use of pre-industrial carbon levels as a baseline against which to calculate natural carbon dioxide levels and to measure more recent changes. The climate in the last one hundred years and more, during our lives and that of our parents and grandparents, has been pretty stable. This time of interglacial warming, which we call the Holocene and which has endured since the end of the last ice age, has been the most stable, climate-wise, in over 100,000 years. But it has been stable only as compared to the ice-age climate preceding it. It has had its ups and downs; conditions have alternated during this time, back and forth, especially in the last several thousand years, between warm and cold. The climate, in a broader sense, is constantly changing.

The official position would have us believe the level of carbon dioxide has remained stable over the course of the Holocene, but this is not so. It has been generally on the increase for the last 7,000 years, and we know temperatures, instead of increasing in line with the official story, have actually declined a couple of degrees since then. If we consider 1850 as a starting date for the Industrial Revolution, that date is also roughly the end-point for that period of climate cooling we call the 'Little Ice Age.' It was five hundred years of the most severe cold weather since the end of the last ice age- that is, since the last big ice age. Knowing, as we do, that carbon levels tend to tag along after temperature changes, we might expect carbon to have declined a little during that period- which it did. Who would think the transition period between an unusually cold and unstable climate that prevailed before 1850 to the modern warming that followed would serve as a good basis- an average of sorts- on which to place something of such obvious importance to us as understanding our changing climate?

Now let's take a closer look at that claim that carbon dioxide levels have gone up in the last half-century to the highest they have been in 700,000 years. One need only look at the carbon record and the process involved in obtaining that data from glacial ice to regard such a statement with high suspicion. Figuring out such things as carbon and temperature levels in layers of ice that are thousands, tens of thousands, and even hundreds of thousands of years old is no simple matter. There is always at least a small degree of uncertainty. It is not so easy as going out and sticking a thermometer or other appropriate measuring instrument in a layer of ice and writing it down.

While lake and seabed cores may be relied upon for more recent measurements, only glacial deposits provide long-term records. The layering in ice deposits is defined by boundaries produced by seasonal changes and other events. Within those layers are gas bubbles containing little samples from the ancient atmosphere when those deposits of ice were at the surface. Now, those gas bubbles do not stay put in their original locations. All gases naturally rise, and, just as they do so in the air, they rise through the ice as well. As time passes, gas bubbles in ice deposits, depending on the ice's density and temperature, will rise at some rate into the layers above. Determining the layer of origin is not easy- it involves some calculation. And to make things even a little more complicated, as you go deeper and farther back in time, that record gets harder and harder to read- and, therefore, less reliable.

As a consequence, long-term records of carbon are most accurately determined for periods of thousands of years. And those changes that occur over shorter periods, like centuries, and especially something so short as a matter of mere decades, such as we are witnessing today, are simply not readable in the climate record.

Another thing to be aware of is that all atmospheric greenhouse gases, including carbon dioxide, have saturation limits; there are limits to how much heat they can retain, and those limits are subject to change. As carbon molecules, or any other greenhouse gas, accumulate in the atmosphere they will eventually reach a point of saturation beyond which they will begin to lose their heat-carrying capacity. Reaching that limit, that capacity declines geometrically- very fast. All of the greenhouse gases are presently well-progressed within their saturation regimes. And this is some good news. It means that although carbon will continue warming as more is added to the air, it rules out a drastic over-heating scenario assumed in projections of a run-away greenhouse effect.

Atmospheric carbon, in fact, was present in much higher amounts in the past. Four hundred million years ago, it was 12 to 16 times its present level. It has gradually decreased, primarily due to long-term biological changes in ecosystems; specifically, increasing amounts of carbon have been incorporated into soils and ocean water through the evolution and growth of plants. Irrespective of these much more elevated levels, however, no run-away greenhouse effect has ever occurred.

Carbon dioxide is good for plants, so in that respect, a little more of it in the air is certainly a good thing. It means more oxygen in the air because plants produce oxygen. This is also a good thing, considering a relatively recent finding that the total atmospheric content of oxygen has declined a little bit. It also means a more plentiful food supply, which is also good- the global population is still increasing.

Most people believe the carbon dioxide theory of climate change simply because it is all they hear. It represents, at best, an oversimplified view, and its role in climate change will eventually be found to be a very minor one. Even though a lot of grant money for scientific research is spent on climate models, their value is only a limited one. They may include a huge amount of data, but they rely too often on a too narrow range of the actual evidence for the global applications where we would want to apply them. If we took that official version and turned it completely inside-out, we would have a more accurate account of what's happening.

The explanation provided here, based on the record of the past, is a little more complicated, but it is a much more viable and realistic explanation than the official story. And once people have all the facts, hopefully, anyway, things will change. We can at least find some consolation in that the finding of innocence on the part of little carbon in the crime lets us humans a little off the hook, too, as the instigators of the whole thing. In the meantime, valuable time is being lost in this pursuit of a people-caused increase of a minor atmospheric gas when we should be spending that time preparing for the changes that are undoubtedly coming.

So the answer to the question of what is wrong with the official story is- just about everything! Why the government and the mainstream media are so busy promoting such a story as an explanation for climate change is perhaps a subject for another day. Carbon dioxide is not the leader in climate change; it does not have the long-term heating capacity in the air that is widely assumed- and there is not enough of it to exert a determining influence on the climate. Carbon does have a useful role to play in our climate scheme of things, but that an element that occupies such a small place in the total climate equation can exert a controlling influence over our planet's climate once you have taken a good and objective look at the facts, is simply not believable.

The temperature/carbon relationship

In the last 100 million years or so, an increase in mountain building has led to higher rates of weathering and a further reduction of carbon in the air as it began to be drawn down at a faster rate by more active atmospheric processes. Pre-industrial levels may have actually been a little lower than 300 ppm- closer to 280 to 290 ppm. Both carbon dioxide and methane are concentrated in the lower atmosphere. Methane levels are even smaller, and unlike carbon, significant variations occur by region, instep with changes in Insolation. But its increase over this period appears even more dramatic, from 700 ppb (parts per billion) before the industrial revolution to 1730 ppb today. For comparison, carbon and methane concentrations were measured at about 200 ppm and 300- 400 ppb during the last glaciation, respectively (Cronin, 1999, pp. 442-443).

A correlation between temperature and greenhouse gases extends far into the past. Following a peak in temperature and carbon dioxide more than 50 million years ago, both have generally trended downward. In the last several million years, as the prevalence of polar ice has increased, an increasing oscillatory nature has characterized the climate- this is especially true for the last 1½ million years or so. Temperatures and greenhouse gases have followed the same trend as the climate has alternated between periods of glacial and interglacial climate. As the climate warmed up from the Little Ice Age, greenhouse gases increased a little more gradually but otherwise in pace with rising temperatures.

The climate record shows that the level of carbon dioxide naturally increases at times of climate warming. It is during such times of warming that the relationship between temperature and carbon is closest. It is evident that at the inception of interglacial periods, temperature increases are followed within 1,000 years by a corresponding increase in carbon.

As the warm period ends and the climate returns to glacial conditions, the correspondence is much weaker; temperatures decline first and are followed much later- several thousand years later- by a decline in carbon (Cronin, 1999, pp. 449-450). The cooler glacial oceans are thought to provide the best long-term sink for drawing carbon down from the air.

The timing of the Younger Dryas event in the glacial cycle is different since it preceded an interglacial warming, and it never reached the severity or the global extent of the Pleistocene Ice Age, but it shows the same relationship. Analyses of deep-sea cores from the Atlantic indicate that temperatures declined 200 years before carbon followed suit; fossil and isotopic evidence from ancient seabeds reveal the same story of climate change. It is reasonable to see atmospheric carbon, methane, and other greenhouse gases collectively as an amplifier at times of warming, as what appears to be occurring today. However, it is also evident that we should not expect their elevated levels to lead the climate into an increasingly warmer future. It is apparent that when temperatures decline, atmospheric conditions can become rapidly reorganized, and they will do so irrespective of the level of atmospheric carbon and other minor greenhouse gases.

Life on Earth is made possible, of course, by the warmth of solar energy. Minor changes occur over time in the amount of energy emitted by the Sun and in the amount of this energy reflected back into space. The Earth's climate, overall, can be seen to be very much a closed system where a relative balance of the total energy is maintained within its global atmospheric envelope, so to speak. Wide variations occur depending on latitude, altitude, and topographical features, but many climate events are understandable simply as a redistribution of atmospheric and surface energy and do not necessarily represent a global change.

Through the looking glass- warming and cooling

In terms of climate, we seem faced with a world of opposites that is mindful of the adventures of Alice in Lewis Carroll's 1871 story, *The*

Looking Glass; a given event in one location or region is often found to be balanced by a similar corresponding event, in opposite sign- through the looking glass- somewhere else. In the mid-90s, as sea levels increased in the subpolar region, significant declines occurred in areas of the Atlantic Current and Gulf Stream (Esselborn, 1999). Sea surface temperature (SST) in the Nordic Seas and the coastal area of Western Greenland tend to occur in opposite sign; as the Nordic Seas began to warm up in the 70s, there was a corresponding decline in Western Greenland. On a broader scale, farther south, when surface waters warm in the eastern tropical Pacific with the onset of an 'El Nino' event- which is brought on by an eastward shift of a large region of warm surface water, usually centered in the western Pacific; storm activity- including hurricanes- declines in the Subtropical Atlantic, as the surface water there cools.

The most important of such events involves the cooling that has occurred in the Stratosphere beginning in the early 1950s. It is interesting to note that just as there is no uniformity in the geographical distribution of the current warming, neither are temperatures in the layers of the atmosphere following a uniform trend. The warming is, in fact, confined to the lower atmosphere- the Troposphere- at the Earth's surface; 10 miles up at the beginning of the thinner upper atmosphere, in the lower Stratosphere, temperatures have cooled down. There is no wide agreement about the processes involved, but the cooling in the Stratosphere is considered to be a natural consequence of the warming at the surface. This mirror image correspondence between lower atmospheric warming and upper cooling appears least intense in the Arctic. The surface warming there has markedly increased in some areas, yet stratospheric temperatures at high latitudes have changed the least. Upper atmospheric temperatures have declined at a greater rate equator-ward, with the strongest contrast with the lower atmosphere occurring in the tropics, where surface warming and stratospheric cooling have been most intense. A rather sudden divergence of the two occurred in 1993, coincident with the beginning of some of the more dramatic changes in the North Atlantic area (Tropospheric & Stratospheric Anomalies, 2012).

So what does this mean? Today's climate debate on global warming refers to increasing temperatures at and near the Earth's surface; temperatures are found to be changeable and vary widely geographically- especially by latitude. These are important points to keep in mind as we continue to investigate the changes that are occurring. The inevitable question that arises, in light of this evidence, especially the upper atmospheric cooling, is how much of the current warming truly represents an overall change in global temperature? And the answer to that question, especially if we are taking that total atmospheric envelope into account, is that there is no warming at all. The true overall global trend is towards cooling.

The Sun rules the day

The dominant influence of the Sun on the Earth's climate is unquestioned. There is no doubt that the orbital influences of Precession, Axial Inclination, and Eccentricity, which produce major changes on the Earth, do so by effecting changes in the distribution of solar radiance. Besides these, cyclical changes in solar energy exert an important and determining influence on global temperatures. Unlike carbon dioxide, which follows a much slower progression, temperature changes can occur unexpectedly fast- sometimes in the oceans and especially the atmosphere.

Changes in solar energy are known to occur over centennial, millennial, and larger periods. Today, satellites keep a constant watch on the Sun, providing daily and even up-to-the-minute reports on developing solar storm activity. Historically, though, an approximately eleven-year cycle of sunspot activity has been the primary determinant for taking a measure of the Sun's radiance. There is an unmistakable correlation between solar radiance and temperature in the climate record. Along with the optimal orbital conditions that prevailed at the beginning of the Holocene, solar radiance was also high then, as compared to today, and historical records show that there were four periods of low, or zero, sunspot activity that correlate with the time of the Little Ice Age. The Sun's radiance increased along with temperatures as the Arctic climate warmed up from the Little

Ice Age. By 1910, temperatures were increasing in many regions, and since then, the Sun's eleven-year cycle has strengthened, along with the warming, with large increases in sunspot numbers beginning in 1947. It is not just the peak years of sunspot activity that are important, however; records from the early 1900s and before show that 1 or 2 years of increased sunspot numbers occurred at times of maximum activity; more recently, clusters of 4 and 5 years occur at each maximum- all of them with relatively high counts (The Sunspot Cycle, 2012).

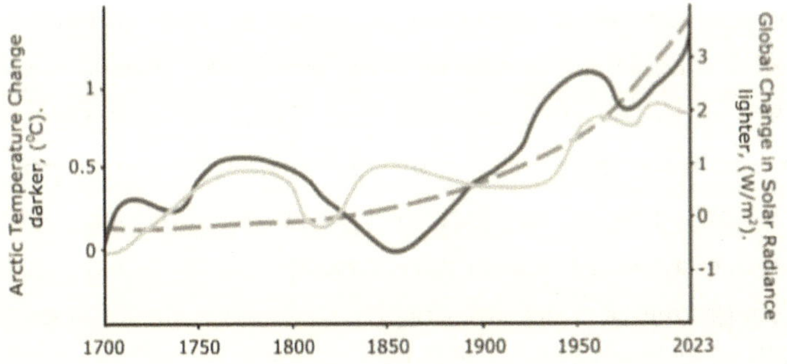

Solar radiance, temperature, and carbon in the Arctic during the late Little Ice Age and modern warming. Broken line represents carbon dioxide. Note their general correspondence- data source: NOAA (Mrugala, 2012).

In perspective, the insolating effect of the increase in solar radiance in the Arctic since the end of the Little Ice Age amounts to about 2 or 3 w/m². It is, of course, the warming that has occurred in the last century that is of primary concern today. In the shorter term, when analyzing individual climate events, the correspondence between solar radiance and temperature is less clear, and the influence of solar energy on climate change is less apparent. For example, a mid-century peak in solar activity included an all-time high in the sunspot cycle in 1958. However, this peak, instead of directly coinciding with a specific warming event, occurred between the early 20th Century warming and the current one that began in the 1970s. Studies of near-surface temperatures from high Arctic latitudes- from tree ring data, for example- show a similar relationship to solar radiance; temperatures there have not, until the last decade or so, reached the levels that occurred in the early 1900s.

More recently, a downward trend has occurred in solar radiance, which may reflect a short-term change; temperatures in the Arctic, as elsewhere, are still generally on the rise. But this downward trend may also lead to a longer-term change toward cooler temperatures on the Earth. Beginning about 1650, during a period called the 'Maunder Minimum,' which characterized some of the coldest temperatures of the Little Ice Age, sunspot activity was at or near zero for over 50 years. Since then- over the past several hundred years- a centennial cycle of lower sunspot numbers and cooler temperatures is evident in the climate record. The early 19th and early 20th centuries were both times of reduced solar activity, and as we enter the early decades of the 21st Century, we should not be surprised to see this pattern repeat itself. Although, recent solar events reflect a rather significant increase instead of the expected down-trend, suggestive of a conjunction of a second cycle of solar activity coming into play- more about this later.

Acceleration of the Hydrological Cycle

Those primary movers of climate change, Precession, Axial Inclination, and Eccentricity, coming together as they did in the Early Holocene, have given humanity a long period of a primarily warm and relatively comfortable climate within which to grow and thrive. Now, some 150 centuries after it all began, we see signs of that warm climate coming to an end. We need to understand the causes exactly bringing about that end. We need to know the signs to look for, and we need to know the effects of the change so that we may see our way forward in this brave new world.

The Earth's axis, since its point of Axial Inclination Maximum, when it was at its greatest angle toward the Sun 7,000 years ago, has since turned one full degree away from that maximum, and Precession passed the halfway marker in its 23,000-year cycle almost 900 years ago. The focus of the Sun's energy and light that began this period of warmth in the north, the point of Precessional Maximum, now favors the Southern Hemisphere. The north-to-south progression of the Sun's warmth on the Earth's surface- of Insolation- has moved from the northern latitudes to the south. And in so doing, it has produced an increasing tropical warming, which in turn is causing an acceleration of the Hydrological Cycle.

As we look at the history of tropical temperatures, we find, as already described in previous chapters, that they have gradually increased since the beginning of the Holocene, which is a natural consequence of the changing focus of the Sun's energy on the Earth's surface over the course of the present precessional cycle. The movement, north-to-south, of this focus of energy is evident in temperature analyses of deep-sea cores taken from many parts of the Atlantic.

Most importantly, the increase in tropical temperatures is producing a strengthening, or acceleration, of the Hydrological Cycle worldwide, which is a naturally occurring result of the warming. Also called the water cycle, this process involves the cycling of moisture through the environment. Water evaporates from the Earth's oceans, seas, and landmasses into the atmosphere and then returns to the surface as precipitation- rain or snow- in an ever-moving and changing system of atmospheric processes. As the warming and evaporation increase, in the tropical oceans in particular, there is an increase in atmospheric moisture and cloudiness and an increasing contrast between tropical and arctic temperatures, accelerating the Earth's natural cycle of equator-to-pole transfer of moisture. Temperatures in the polar regions have declined more than they have increased in the tropics. More than a temperature change, it is this contrast between tropical and polar temperatures that is the actual driver bringing about the ever-quickening pace in this tropic-to-pole movement of air, along with a lot of heat and moisture. And this is having a very significant and increasing impact on climate.

This warming in tropical latitudes includes a large area of the Earth's surface. Rising temperatures have increased the evaporation rate- especially important, of course, in the tropical oceans. More evaporation means more moisture in the air. An increase in atmospheric moisture- of water vapor- and cloudiness has been known since the early 1900s. A resulting increase in heat and water is particularly evident in the Northern Hemisphere, in the northern Arctic region, due to a sharper contrast in tropic to pole temperatures.

In the area of most concern, centered on about 65° N, in the far North Atlantic- the general area of Iceland- precipitation has increased in the last few decades by at least 30%. River run-off from surrounding landmasses into the Arctic Ocean is correspondingly up at least 20%. Moreover, all this added water has to go somewhere. And in the Arctic, with the ocean nearly surrounded by land, there is the Bering Straight, between Siberia and Alaska, but the water flow there is the other way- from the Pacific into the Arctic Ocean. The only outlet for all this water is in the eastern Arctic: south, through the Canadian Archipelago west of Greenland, and east of Greenland through the Nordic Seas- all of it ends up in the North Atlantic.

The effect of an acceleration of the Hydrological Cycle is to extend the influence of increasing tropical temperatures and moisture into the middle, and perhaps even more importantly, high latitudes, in both hemispheres. Due to the greater temperature difference between Equatorial and, in particular, sea surface temperatures (SST) in northern latitudes- as illustrated in Chapter 6- its impact is most pronounced in the Arctic.

Plainly said, the warming in the Arctic is a direct result of the tropical warming. Carried there, along with the extra water, by way of the quickening Equator-to-pole flow of air. A continuing increase in the contrast between tropical/polar temperatures can only amplify the effect.

And there is even more to it than the addition of water and heat to the polar regions. As it turns out, water and carbon dioxide are not the only greenhouse gases on the rise. Because of the strengthening Hydrological Cycle, the levels of all atmospheric greenhouse gases, including methane and nitrous oxide, are going up. This is all due to natural processes; none of it can be blamed on human activity. One need not look beyond this single overriding factor to find ample reason for increasing amounts of greenhouse gases. A fact that should be taken into account- particularly with respect to carbon dioxide.

Essentially, what we've got is an acceleration of atmospheric processes- an acceleration of everything! Along with the quickening of water cycling through the environment goes everything else, bringing on increasingly active atmospheric conditions into the mid-latitudes and an acceleration of generally deteriorating conditions- particularly farther north.

The resulting accumulation of moisture in the atmosphere is directly responsible for many of the problems we face today. A couple of familiar clichés come to mind with this: all this water in the air has to go somewhere, right? Like everything else, 'what goes up must come down.' And, nowadays, 'when it rains, it pours'- literally. Those days we have known of gentle summer rains, all too often, are replaced with downpours.

Consequently, there is a great increase in flooding. Storms tend to be bigger, stronger, and more frequent- more powerful hurricanes and tornadoes. The importance of this acceleration of atmospheric conditions in today's climate is self-evident globally and in the North Atlantic region.

As evaporation increases, there is a corresponding decrease in surface moisture in land areas and an expansion of arid conditions. A drying climate is a very familiar problem in many regions now; the portion of the Earth's surface subject to severe drought has more than doubled from 10 or 15% in the early 1970s, perhaps exceeding 30% today, with accompanying decreases in freshwater supplies and a sharp increase in wildfires in many areas (Drought's Growing Reach, 2005).

Increasing drought conditions from Africa into Eastern Asia are attributed more to an increased frequency of El Nino and- corresponding opposite- La Nina events, with the large mass of warm tropical surface water mentioned earlier, and which is usually associated with western Asia, increasingly migrating back and forth across the Pacific.

But the effect is global; a drying of lakes and streams- and even rivers - and the soils needed for agriculture is particularly noticeable in the mid to high latitudes in the Northern Hemisphere. The overriding cause is this same strengthening of the Hydrological Cycle. These changes are unprecedented within a historical context. They have occurred many times in the distant past, as evidenced in the geological and fossil record- the climate record- but they are new to us.

A narrowing of daytime temperature ranges (DTR) is especially evident in the Northern Hemisphere beginning in the early 1900s, concurrent with the overall warming trend. In most areas, this involves an increase of temperatures within the lower range of daytime temperature averages; in other words, a warming up of the cooler times of the day. This suggests that some of what usually falls under the description of 'global warming' may be more accurately described, not necessarily as always warmer temperatures, but simply as more of it.

The most notable exception is the Nordic Seas region, where temperatures have not followed the warming characteristic of the rest of the Arctic. There, the higher (maximum) daytime temperatures change the most, trending downward, but not at a rate to indicate a significant cooling trend. Another, more visible, aspect of this effect is an apparent migration of the daytime temperature high towards the later afternoon. These changes in the DTR are coming about due to increasing atmospheric moisture and strengthening of the Hydrological Cycle (Tuomenvirta, 2000).

Warmer air is moister air, and vice versa; just as adding moisture to the air in your house with a humidifier in winter makes it warmer, increasing atmospheric moisture, along with the strengthening Hydrological Cycle, is an inherent aspect of the tropical warming. Unlike carbon dioxide, which changes very slowly, the moisture content in the atmosphere follows very closely with changing temperature.

The accelerating Hydrological Cycle may ultimately exert the determining influence on the Earth's climate. As there is a return to glacial conditions at the end of an interglacial period, atmospheric moisture levels and temperatures can change rapidly (Adams, 1999). A tropical warming and a strengthened Hydrological Cycle are found in the climate record to be dominant climate characteristics toward the end of an interglacial period. Studies of the last glacial period show that small changes in the Hydrological Cycle can induce large and rapid shifts in the Earth's climate; that under such conditions, the climate can change rather abruptly to a completely different climate state (Clark, 2002; Taylor, 1999; Vörösmarty, 2001). We should consider these changes we witness today to be so significant as to require our absolute and undivided attention.

Solar and Galactic Influences

The Universe is a vast, mysterious, and fascinating place, with billions of rotating galaxies, each containing billions of gaseous nebula, rotating stars, and their planetary systems- full of organic and inorganic matter and the possibilities for life. The largest things we know of have a lot in common with the smallest things we know. Orbiting Stars, planets, and their moons are connected by the same electromagnetic fields of influence as atoms with their orbiting subatomic particles. The Universe, therefore, is full of the same electromagnetic forces, pushing and pulling on one another. From the massive galactic centers to the very core of our planet, electromagnetics is everywhere. All matter is a form of energy. All that we know, from the smallest to the largest things that exist, are made up of the same electromagnetic particles and fields.

There is no doubt that electromagnetics exists within everything, and all things are connected to it and influenced by it in some way. Changes in that same energy, coming from the Sun and beyond our little system of Sun and planets- from the very center of the Milky Way Galaxy- exert influence over everything else.

Modern electronic communications are susceptible to occasional bursts of plasma emitted by the Sun, which we call CME- coronal mass ejections. The largest such event on record, where the Earth was hit by one of these, occurred in 1859- the Carrington Event. There were no satellites or electronics then, but there was an extensive telegraphic communications system, which experienced many explosions and burned lines. There is archaeological evidence of much larger similar events. Ancient monuments show the appearance of wide-scale burning of the surface, having occurred on the order of thousands of years ago. These are best explainable as having been produced by one or more, likely much more prominent, such events.

One thing we know for sure is that our Sun is anything but predictable! Other effects on the Earth's surface that such bursts of solar activity may influence include increased earthquake and volcanic activity. There is a general correlation evident between the occurrence of geological events on the Earth and these bursts of solar matter and energy. Increased solar activity also produces more polar auroral displays- northern and southern lights. Today, they become more brilliant and extend more often into the lower latitudes- a more visible sign of an increasingly active, more energized solar environment.

However, the Earth's electromagnetic field that protects us from harmful incoming radiation is now weakening. Its magnetic poles, often thought of as opposite one another, like a pole magnet, are increasingly migrating equator-ward, raising concern of a possible reversal that would leave the planet's surface more exposed for some undetermined period of time. Although the dramatic events surrounding the climate changes are occurring principally in the North Atlantic area, much of the weakening in the magnetic field is centered in the South Atlantic and at the South Pole itself (Kakad, 2022). All the while, there is a not-too-surprising measured increase in cosmic radiation- the primary source of which would be the Galactic Center.

So, we have two possible influences to consider- either solar or galactic- to which we might attribute this decline in the strength of our planet's electromagnetic field and the changes in the positions of its poles. The Sun, we know, rules much of the time out here in our little corner near the edge of the Galaxy. The Galaxy is a big place; among the many billions of stars (suns) and planets besides our own, its center- the Galactic Core- is by far the greatest of entities. It is, for us, the single predominating source of energy- of light.

A look at what is happening elsewhere in our solar system and beyond may help to clue us in on what's going on. As it happens, it isn't just the Earth; nearly all the planets show signs of relatively recent change. Besides Earth, the electromagnetic field of Jupiter also is weakening. Instruments on the Martian surface are, for the first time, picking up signs of seismic activity. Several planets, including the Earth, are experiencing slight accelerations in their annual solar revolutions. All the outer planets show signs of increased atmospheric changes; Jupiter's red spot is no longer red, and Neptune is experiencing significant changes in cloud formation.

Looking farther afield, some nearby stars appear to be becoming more active. The giant red star Betelgeuse, 650 light-years away in the constellation Orion- considered nearby in terms of stellar distances- is a subject of concern due to its recently increased activity (Castelvecchi, 2020). Betelgeuse is nearing the end of its stellar lifetime, having nearly exhausted its hydrogen/helium content, and the worry is that it will end in a colossal supernova explosion. That end may be a thousand or more years away, or it may happen tomorrow- and it is too close to us not to have an effect. At any rate, attributing all of this to just our Sun would require a great stretch of the imagination.

Of course, we are all accustomed to short-term changes in solar energy. Modern-day satellites keep us posted on the Sun's activity and alert us when a magnetic storm on the solar surface sends out a burst of plasma and radiation in our direction. Like everything else, the Sun's energy level goes through cycles of change, continually increasing and decreasing over time; the one best known is the eleven-year sunspot cycle- described earlier. Sunspots are evidence of this cycle of increasing and decreasing storm activity. There is a record of their occurrence going back to about 1750. The coldest part of the Little Ice Age, which coincided with a remarkably long period of reduced sunspot activity, is suspected to be a low point in one such cycle of change in solar radiation.

There is an apparent correlation between periodic changes in solar and cosmic energy and primary events on Earth associated with major climate change. Of particular interest in this respect are solar cycles. Besides the sunspot cycle, a 1500-year cycle, as described in Chapter 6, is evident in the climate record during the Holocene and the Dryas period of glacial climate that preceded it. It seems that every single climate event identified in this analysis, of both warming and cooling, beginning when temperatures first rose above glacial levels about 15,000 years ago, finds a close correspondence with this cycle. The beginning of the Holocene was a time of relatively high solar activity, corresponding with an apparent 12,000-year cycle that can be traced back well into the glacial era. And within that, there is an apparent, although less intense, 6,000-year cycle of change that corresponds with the Mid-Holocene Optimum. This means that today, some 11,500 years after the beginning of the Holocene, we should not be surprised to see an increasingly active Sun. Which we are presently seeing, even though it is also presently at a point of expectedly reduced activity in a 100-year-long centennial cycle. There is a linkage between such changes in solar radiance and its effect on both decadal and long-term aspects of climate, including the AMOC (Ye, 2023).

What's perhaps more interesting is that the early Holocene may also have been a time of increased activity in a wider Galactic cycle of change. Besides the solar changes, galactic energies are in close correspondence with the important changes that are the subject of this analysis- they are all happening at the same time. Just as they have played a part in the big changes of the early Holocene, we should not be surprised to find more, both larger and smaller, cycles of overlapping change in solar radiation. And we should not be surprised to find both solar and galactic sources causal factors in the accelerated- energized- climate conditions we see today. As though we don't already have enough on our plates with what's going on here on the Earth's surface, and particularly because of our apparent place in this 12,000-year cycle, we might just find ourselves in the coming years also witness to some dramatic and unprecedented solar/ galactic events.

We have learned a thing or two about how climate changes on the Earth, but about the part that solar and galactic forces play in it all, we are just learners. That these extra-terrestrial events hold the ultimate trump card with regard to climate conditions on the Earth and over every single aspect of the climate we have examined in this analysis seems an unavoidable conclusion.

When you think about it, how else could it be? What other forces can set a planet in motion, determine its physical properties, the angle of its axis- the size and shape of its orbit among its fellow planets and the stars? The Sun will be the local favorite in terms of influence, but we should not bet a lot of money against that Galactic big boy whenever it's in town.

PART FOUR

THRESHOLD TO A BRAVE NEW WORLD

10

A CLIMATE IN TRANSITION

Signs of change- a new cooling trend

A transition from one climate state to another occurs due to changes in oceanic and atmospheric circulation originating in the North Atlantic and Subpolar Seas. The Atlantic Current, both at the surface and depth, and the associated convection regions in the Subpolar Seas comprise the most critical link in ocean circulation and are particularly sensitive to change. As described in Chapter 6, changes in glaciation associated with major climate events are found to begin along the latitude of 65° N, directly across southern Greenland and northern Iceland (Adams, 1999). The apparent place where ice ages and interglacial warm periods begin and end- it is the first and best place to look for clues to major trends.

Three basic modes of ocean circulation have been identified for this region, and each one is associated with a specific climate state, which can generally be seen to also apply on a broader global scale. The present mode occurs during interglacial climate and produces the warmest and most humid conditions, during which the NAO is in its most strongly positive phase. At the height of this mode, as during the Early Holocene, the northern influence of the Atlantic Current and the strength of the Deep Current are at maximum, and convection, especially in the Nordic Seas, reaches its greatest extent and depth. A second mode, which prevails during much of glacial climate, is characterized by colder, drier, dustier,

115

and more active atmospheric conditions. In this state, convection is significantly weakened in the Nordic Seas, and the flow of the Deep Current is much diminished. Convection continues in the subpolar region, however, even at depth, while the influence of Antarctic water becomes widespread in the North Atlantic Basin. The Atlantic Current continues to flow, although at a perhaps much-weakened rate, with little influence north of the Greenland-Scotland Ridge. Another less prevalent mode corresponds with the coldest of glacial conditions and large ice-rafting events in the North Atlantic, which may involve the total cessation of the Atlantic Current in northern regions (Taylor, 1999; How Fast Did Climate Change, 2001).

Variations are known to occur within each of these modes of circulation, and today's conditions find us still enjoying an interglacial climate. However, as time goes on and we get further and further away from the early Holocene optimal conditions that produced it, the climate will increasingly favor a return to its usual glacial state. It might be interesting to note that Antarctic water is also present in the North Atlantic during an interglacial climate. It is there today, at the very bottom of the ocean basin, and flows into the region through a smaller current near the surface, directly below the much greater north-bound Atlantic Current.

Looking at previous interglacial periods, it is evident that by the time Precession had returned to its minimum position in the Northern Hemisphere, the precise point of which it passed in the present cycle almost a thousand years ago, the climate was nearing a point of transition to colder conditions. No one knows, of course, precisely when the present warm period will end; it occurs a little differently every time- it could still be a long time from now.

But the events observed in the North Atlantic and the Nordic Seas in the last several decades suggest a much more rapid progression of change. The large and increasing influence of Arctic water- most especially the appearance of the Atlantic Cold Blob; the changes in temperature and salinity, the weakening and increasingly higher frequency changes in atmospheric conditions, most exemplified by the weakening of the

Jet Streams in recent years; the dramatic down-turn of the convection process, the decline of the Subpolar Gyre, and the decreasing strength of the Atlantic Current- both at the surface and at depth- are all associated with times of major change in the climate record.

An ancient priest or scholar, who may have been watching such things 5,000 years ago as the Early Holocene climate was coming to an end, might have noticed some similar changes. The climate at that time cooled down by a couple of degrees in the far north, while in the lower latitudes, conditions became drier, and the NAO changed to a weaker, more negative circulation pattern. At the time of the Little Ice Age, just a few hundred years ago, northern conditions would have been seen to be the most severe since the end of the Younger Dryas cooling, with the NAO correspondingly negative.

A new and persistent NAO negative phase has today taken hold in the North Atlantic, and a large southbound inflow of Arctic water is now decades in the making (Calvin, 1998). But it is not just in the North Atlantic that the climate system is slowing down; as mentioned in Chapter 2, The Arctic Oscillation (AO), which governs weather events for all of the Northern Hemisphere, has also moved into a generally weaker phase. At about the same time- the late 90s- a similar event, along with similar relative sea-level changes, occurred in the North Pacific- called the Pacific Decadal Oscillation (PDO) (Volkov & van Aken, 2005). And prior to that, in the South Pacific, 1976/77, the Southern (Darwin) Oscillation- an aspect of the PDO- also changed to a negative phase. Are we observing the beginnings of such a large climate change today?

To the list of changes, we must add at least two more influences that are particularly global in reach; the current tropical warming and resulting changes in the Hydrological Cycle appear, in this analysis, to be most important in driving climate change, both in the North Atlantic area and worldwide. Periods of warming and changes in this essential global cycle are closely correlated with large climate change, as are the events we see in the North Atlantic and Nordic Seas- all have these same global influences as their primary cause. It is this combination of these regional changes as well as the global ones that is the concern in this analysis.

The changes in the NAO are as telling as anything of the importance of the events occurring in this region. The decline that has prevailed since the mid-90s is truly unprecedented for the period of the historical record. The new record lows in the winter index demonstrate the increasing oscillation of weather conditions, while the annual and especially summer indexes clearly show the record negative shift that has prevailed since 1996. It is evident from the climate record that summer conditions weakened toward the end of the Eemian Interglacial Period, a symptom clearly present now in this data for the NAO. It is interesting to note that in the Antarctic, where some record-setting cold winter conditions have recently been recorded, there has also been a record of a profound cooling trend in summer conditions for much of the time since 1980. These are some things to take into account as we examine the evidence of our place in the overall scheme of things.

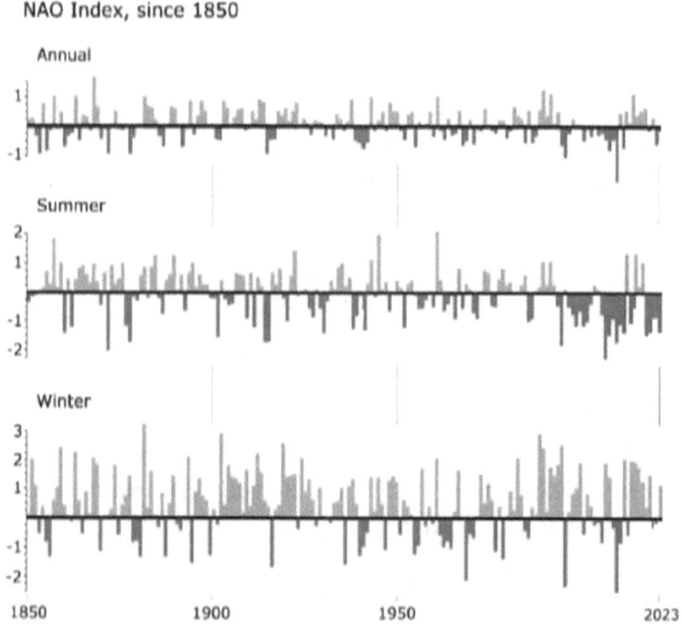

The summer index is for June- September, and the winter index for December- March (Hurrell, 1995; Jones, 1997; North et al., 2011). A numerical index (and included notes) for 1990-2023 are at the end of the text.

It may seem somewhat paradoxical that a warming, like today, could usher in a period of climate cooling, yet evidence shows that the climate has progressed just this way in the past (Eiriksson, 2006). Clearly defined glacial deposits that reveal conditions at the very end of the Eemian have not yet been found, but such a correspondence between warming and cooling events is evident at that time. While the Eemian was slightly warmer, with a greater degree of seasonality and perhaps wider swings in temperature, like the Holocene, it was characterized by an early period of climate optimum leading to a mid-period warming followed by a change to a cooler climate. As a similar warming got underway in the mid-Holocene, changes in atmospheric conditions and ocean circulation began, creating a variable climate situation similar to that which exists today, and an alternating cycle of warm and cold periods continued to play a primary role in the late Holocene cooling and neo-glaciation that followed- and which has continued, more or less, until modern times.

Likewise, both the Roman and Medieval warm periods were followed by abrupt cooling events, most notably, and recently, the Little Ice Age (Drange, 2005). Both the Younger Dryas and the 8,200 yr. Cold events are attributed to large freshwater increases, and both were brought on by periods of Arctic warming and melting. On a more immediate basis, the same relationship appears to apply to annual changes in weather, of which the winter of 1995/96 and the one preceding it is one prominent example, where strongly negative shifts in the NAO tend to be preceded by equally strong positive ones.

This evidence all leads us to the same conclusion: that the present course of events in the North Atlantic and the Nordic Seas will ultimately lead to a cooling. Beginning in 1996, the NAO and summer conditions generally are in decline, and precipitation there is more than sufficient to begin a new glacial expansion. The warming in the Arctic is still dominant, but in this area, there has not been a significant trend toward warming or cooling. That is, until 2010, with the beginning of the new dominance of the Cold Blob.

In the North Atlantic and Nordic Seas, air and sea temperatures have begun to decline since the climate shift in 2010. Arctic sea ice volume, in decline since the mid-nineties, after 2010 has nearly leveled off. Coastal glaciers around Greenland and Iceland, although still melting, do so at half the rate they did before 2010 (Noël, 2022). Moreover, Greenland's largest glacier, Jakobshavn, located on its southwest coast, since 2012, has experienced significant new growth. Europe, being generally downwind from much of this, has escaped a lot of the cold temperatures so far, but storm activity there, especially in winter, has been becoming stronger and more frequent since 2010- sometimes with near hurricane-force winds.

Does this mean the dreaded decades-long warming has come to an end? Time will tell. This new cooling is attributed to the presence of our Atlantic Cold Blob. Predictably, the warming will sooner or later fall into defeat, but it may have gotten a bit of a bad rap in all of this. Right or wrong, it is, after all, blamed left and right for just about everything that goes wrong. Either way, you had better believe that Blob is no hero. Of course, it may just go away someday, and a few more of them may come and go between now and the end of the Holocene warming. But it might still turn out to be the monster in our continuing story of climate change.

The rapid pace during a transition event

This is not just another late Holocene warm period. This time around, there are some striking differences. The vast expansion of Arctic water and the corresponding weakening of the Atlantic Current- the loss of its great northern arm into the Nordic Seas: These things have not occurred since the end of the Eemian. This is compelling evidence of a significant shift in the water balance that, in the past, has led to important and long-term change. We have learned that such events have been the key to initiating a climate transition worldwide.

It is perhaps the most striking revelation of modern climatological research that while the Earth's slowly changing orbital conditions ultimately determine the course of climate, an actual transition between climate states does not follow this same long-time regimen but occurs rather quickly in response to changing conditions on the surface. The reflectivity, especially of snow cover, of the large span of landmasses in the Northern Hemisphere plays an important role, as does the acceleration of the Hydrological Cycle, which has increased more than just a little over the last several decades. The North Atlantic is the most critical single region with regard to changing surface conditions and of primary concern in global ocean circulation. Such transitions represent the most extreme manifestations of change in the climate record: a complete transition to a new climate state, formerly thought to take hundreds or even thousands of years, with respect to temperature change, is now known to occur in only a matter of decades. Warming events at the beginning of interglacial periods are the most abrupt of large changes; at the dawn of the Holocene, temperatures in central Greenland increased by 7° C (12.5° F) or more in just a few decades, and perhaps 10° C (18° F) in 50 years- with much of the change occurring in only a decade or less. The inception of the Bolling-Allerod warming before that, as temperatures first began to rise above glacial levels, required only 20 years (Taylor, 1999; How Fast Did Climate Change, 2001).

The change characteristically occurs more gradually when the climate returns to glacial conditions. The Younger Dryas event was felt principally in the North Atlantic, Nordic Seas, and surrounding land areas. It did not reach the global extent that a glacial/interglacial transition does, but it is the most recent of large cooling events. Similar to the early Holocene transition to warming, its initial change to cold conditions was completed in a step-wise fashion over a period of about 45 years. The climate oscillated several times during this change between warm and cold conditions in approximate five-year increments before settling into a glacial climate (Broecker, 1997: *What If the Conveyor Were to Shut Down,* 1999; Calvin, 1991).

There are no written accounts of a transition event- the historical record goes back only a little over 5,000 years- but the essential elements involved are generally understood. Many of the changes that we are observing today; the expanding drought conditions, the increased incidence of dust storms and wildfires, as well as the increased severity of winter storms, hurricanes, tornadoes, and flooding, are all generally attributed to global warming or identified as part of the cyclical nature of climate that we have all grown accustomed to. But these are also found to be dominant climate characteristics in the expectedly more active atmospheric conditions during glacial times, and it is certainly conceivable that they could be early signs of a climate transition.

During such times, conditions in the North Atlantic region and beyond are likely to become increasingly sensitive to environmental changes; seemingly small changes in the Subpolar Seas, perhaps in the water balance or the Hydrological Cycle, may be a tipping point that will initiate changes worldwide. A reconfiguration of ocean currents is a definite signal that the climate has entered into a transition, which we are already seeing some of, and which may quickly bring us to a threshold beyond which atmospheric circulation will become reorganized extremely rapidly. And there are definite signs of that now with the weakening of the Jet Streams.

If the winter of 1995/96 is an example of what is to come, then we can expect that winters, in particular, will be much more severe with large and rapid changes in weather. We can expect extremes of both cold and warmth. Large geographical areas in the early months of '96 saw a number of extreme weather events. Intense cold and heavy snows prevailed in broad, especially northern areas, with almost summertime temperatures and flooding occurring at the same time elsewhere and perhaps only a few hundred kilometers away (Special Climate Summary, 1996).

As such a transition progresses, summers will become shorter-something already apparent in many areas. In temperate regions, there will be a loss of seasonality; it will eventually become more difficult to distinguish one season from another as winter conditions increasingly extends its reach into the rest of the year. Temperatures will become colder worldwide, with successive winters of unusually cold weather in the mid to high latitudes- colder than the coldest winters of the 20th-Century. As a glacial climate settles in, we can expect snow and ice to increase significantly in both polar regions and perhaps in the Arctic in particular, and glaciation to expand dramatically worldwide. We can expect temperatures to eventually return to something similar to those of the last ice age, when air temperatures in Greenland were as much as 16° C (29° F) colder than today, with Antarctica and tropical regions being some 10° and 4° C (18° and 7° F) colder, respectively; and with temperature changes in the mid-latitudes- the temperate regions- somewhere in between (Broecker, 1999).

Extinctions and biomass burning

Large reductions, and also extinctions, of plant and animal life are evident in the climate record, particularly in the north, at the onset of all of the most recent climate transitions, as well as at the beginning of the Bolling-Allerod warming, and again at the onset of the Younger Dryas cooling event. And then again as the climate finally warmed to interglacial conditions at the beginning of the Holocene. These events are very likely coincident with every large climate change. As the Pleistocene Ice Age ended nearly 15,000 years ago, extinctions occurred across Europe, Asia, Africa, and perhaps more in Australia and North America. The large land animals that occupied the glacial and tundra environments, the mastodons, mammoths, and saber-toothed cats, all disappeared. As the Younger Dryas cold ended and the climate warmed again, at 11,000 BP, more extinctions occurred in South America (Elias & Schreve, 2007).

Today, environmental stresses are increasingly evident. Significant declines in numbers and species diversity are evident in many plant, animal, and insect groups. These are widely apparent in terrestrial as well as oceanic ecosystems. Among the most notable is the reduction in the abundance and diversity of bees. Locally and regionally, the change is attributed to increased diseases, habitat loss, or warming temperatures. All of these conditions must contribute to it, but the decline is not just local or regional. It is occurring worldwide and creating a 'pollination crisis,' as many vegetable and fruit crops, as well as other important plants, depend upon pollination by bees and other insects. It brings to mind the quote attributed to Albert Einstein, to the effect that if bees were to disappear totally, pollination of plants would cease, and then humanity too would become extinct.

Declines in other groups, such as amphibians and mammals, are also worldwide; this is a subject of deep concern and not well understood. Longer-term changes evident in the North Atlantic region are more attributable to changes in climate. Analysis of core samples taken from ocean-floor sediments reveals a decline in the total number of sea-surface dwelling microscopic plant and animal species- sea plankton - over the last couple of thousand years. And perhaps not surprisingly, the decline has occurred principally in warm-water-loving species (Witak, 2005).

This brings to mind the widespread replacement of those warm-water species with cold-water type plankton that occurred in the Arctic in the winter of 1996, described in chapters 1 and 3. The warm water species of plankton referred to here have been dominant in the Arctic for the last 10,000 years- since the climate first warmed up to today's temperatures. This is not a good sign for the continuation of a warm interglacial climate; it is instead a sign of change toward colder conditions. A decline in northern tree species that has occurred since the end of the mid-Holocene is also attributable to changes in climate. These declines in numbers and species are reminiscent of the reductions and extinctions that are evident at times of large climate change in the past.

An unmistakable correlation is evident between biomass burning that occurred at the onset of the Younger Dryas, in particular, and the increased drying of soils and marked increased incidence of wildfires occurring today in many parts of the world (Cronin, 1999). Desert regions across North Africa and into southern Asia continue to expand, and other areas, such as eastern Australia and the western U. S., are experiencing significant increases in the length of the dry season and the incidence of wildfires as compared to 1970 (Hotz, 2006).

Environmental stresses and diseases are increasingly affecting plant, animal, and insect species worldwide, and particularly in light of the global experience with the COVID-19 virus, we must not be lulled into believing that this will not have a profound effect on human populations as well. The entire time of the Little Ice Age was plagued with disease and famine, beginning in about 1350 with the Black Death, which claimed many millions of lives until its very end. It is in times of climate cooling, not warming that the spread of sickness and disease becomes most prevalent.

These are some facts that we should take into account as we evaluate our place today in the scheme of things. The onset of glacial conditions will produce many difficulties, and it is not just the cold that we have to worry about. We have an inherent interest in the survival of the other forms of life on this planet if we are to secure that of our own.

The next climate shift

So now we are all wondering when the next big change will happen and what it will be like. The first went largely unnoticed, occurring as it did in the deep sea, and the second one, especially the broad geographical scale it attained, has caught just about everyone by surprise. The Earth's climate is complex, and no one knows precisely how events will unfold. But when the warming of the Holocene finally comes to an end, if it progresses the way major cooling events have in the past, temperatures will begin to decline in the climate-sensitive region of the North Atlantic and Nordic Seas and move into coastal areas of northwestern Europe and northeastern North America. From there, it will gradually extend its influence across large areas of the Northern Hemisphere and beyond.

There is a new cooling trend in the region, attributed to the persistent presence of the Atlantic Cold Blob. However, the main event will directly involve the Atlantic Current and what happens with it in the North Atlantic. During times in the past when glacial conditions prevailed, that essential overturning process that operates primarily in the Greenland Sea during interglacial warm periods is essentially shut down and becomes re-established in the North Atlantic. When the Atlantic link in the system of ocean currents fails, as it does at the end of an interglacial period, it always starts again farther south. The global circulation of ocean water is a natural, ongoing, and inherent process that has gone on since there were oceans to circulate- driven by the Earth's rotation on its axis.

At no time until now, since the beginning of the Holocene, has the great Atlantic Current ever deviated from its northward trek. Until today, it has never given up its dominant place through the northern seas. Even as the critical ocean water overturning machine in the Greenland Sea grew silent, Atlantic water, although weakened, continued in its steady course. But now, much has changed. The Atlantic Current has assumed a more secondary role; it must now give its Arctic antagonist a wide birth. Today, it takes the long way around towards its destination in the Nordic Seas. Directly in its path stands a giant volume of cold, fresh Arctic water. At no time until now, during the Holocene interglacial period, has Arctic water held such a pervasive presence in the North Atlantic.

Today, coming north into the Subpolar Gyre, head-on against the giant Cold Blob, the Atlantic Current, though continually weakened, still perseveres. However, it soon loses some strength as portions of its west and east flanks break away from the primary current. It spreads westward and eastward, turning back south, increasingly slipping deep into the ocean (Bryden, 2020).

This break-up and recirculation has been going on for a long time- since at least 2004- (Bryden, 2005). And there is more and more of it with each successive change. After the big change in 2010, it is evident that it plays an increasingly important role in the weakening of the current going north and its reduced heating capacity. It just might be the beginning of a reorganization that will lead to the re-establishment of its overturning process- lost mainly with the mid-nineties big down-turn in the Greenland and Labrador Seas.

The next climate shift may be strike three for the Atlantic Current, and we don't know if there will be the opportunity for another inning. It will likely come on as quickly as the first two. And it might be tens of thousands of years before there is the chance for a new game.

More than ten years ago, as preparations for the publication of this book's first edition were being made, it was taken for granted that hundreds of years might elapse before the occurrence of some of the events that we're seeing today. An anticipated major cooling is now a reality in the northern seas that in time could possibly propagate into a global transition event. Important atmospheric changes, like in the Jet Streams, are more immediate signs of equally important changes in ocean circulation- already underway. Some are known in the climate record to precede by just a decade or so a significant global change.

If things continue as they have been going, the next shift will include an even larger increase in recirculating Atlantic water and another correspondingly reduced volume of Atlantic water traveling on to the Nordic Seas and beyond. On the day when there is a much reduced current to warm the lands and seas farther north, a cold Arctic climate will quickly descend to take its place.

With each change in the northern seas, the direction in which the climate is taking us becomes ever more clear. There will be a day when Atlantic water coming north into the Subpolar North Atlantic, perhaps along the southern margins of some new winter sea ice formations, will finally complete the connection with those deep ocean currents moving back south, and there will be a much reduced current traveling on north. The global link will have been satisfied; the continuation of the world's major ocean currents assured. No longer then will Atlantic water warm the lands farther north. The northern seas will quickly become quieter and much more cold. The end of the warmth of the Holocene will finally have come. A new era of glaciation will have begun. And when that happens, we are going to want to be prepared.

We have today placed ourselves in a high-stakes gamble with Mother Nature. We are newcomers to some climate events that have not been seen for a very long time and which Nature has played out many times before. We should realize by now that carbon is not the high card that has been assumed; it has obscured our view of far more powerful forces of change.

We are witness to changes of which we have only limited knowledge, both in the Arctic and globally- but we have learned a few things. Every important aspect of the North Atlantic climate is experiencing great change. The warm, humid, green world of the early Holocene was only possible due to a vigorous over-turning of ocean water in the northern seas. Today, that critical circulation process is mainly gone. The warm water boundaries, having been pushed ever southward, have been drawn and re-drawn. Arctic water now dominates in the North Atlantic. Our line in the sand, so to speak, now extends into the middle latitudes. We should know by now that when it comes to the climate, Nature holds all, or nearly all, of the cards. We must learn to adjust to changing climate conditions- and begin working with Nature, instead of against it.

11

LIVING IN A GLACIAL WORLD

Change within the span of a human lifetime

For all the study and analysis accomplished and all the preparations achieved, the threshold that will lead to a new glacial climate will likely be difficult to recognize until it is finally experienced. We can recognize, however, that some beginning events have occurred- some of them hundreds and even thousands of years ago. And it should be apparent to us now that by the winter of 1996, an actual transition event was underway.

We need not worry about a rapid spreading of ice across the countryside, like that portrayed in a Hollywood movie. However, as conditions exist now, those living in large cities and other areas in far northern latitudes are especially ill-prepared for the cold and stormy winters that will undoubtedly prevail. In this analysis, we have seen the evidence and effects of significant changes in the North Atlantic region and how quickly they can come about, as evidenced by the sometimes extensive and rapid atmospheric and oceanic events. The ten-day void in the Deep Current in November of 2004 is one example of an abrupt and unexpected change. A critical part of the climate system- one day, it was there, and the next day, it was gone. How will we deal with similar events in the future? The appearance of the Blob after 2010 is visually and geographically a much bigger event and the most widely known. It took several years to develop, but knowing days, weeks, or even years beforehand might not be enough time to prepare for some extended periods of storms and extreme winter cold that will prevail. Although hardly noticed by anyone, the mid-nineties reduction of the North Atlantic's circulation system was equally significant. All of these events are important steps toward a switch to glacial conditions. There will undoubtedly be changes in the future that will be even a lot bigger than anything we have seen so far and may occur just as quickly.

This is not a prediction of a doomsday event, but it is about an event or a series of events that will affect everyone everywhere; all past and present troubles of humanity will likely pale in comparison. We cannot expect the climate to yield to any attempts at human regulation, and it may be a long while before climate models or computer programs can accurately project the course the climate will take. For the foreseeable future, life on Earth will have to adjust to changing conditions, and to some extent, those adjustments will have to be made as conditions unfold- there, no doubt, will be some surprises along the way. Nevertheless, we have a pretty good idea of how this is all going to play out, and we cannot afford to wait until we are at that threshold, with a cold and dry climate bearing down upon us to act for considerable time and effort will be required to make preparations.

Should such a climate emergency begin today, energy sources and supplies would be insufficient to meet societal needs. As practiced now, agriculture could not keep up with the food requirements of a hungry world and could not be reorganized quickly enough to avoid significant shortages. Mass migrations would likely occur, perhaps numbering in the hundreds of millions, from the northern cold to a less than rosy world farther south where broad regions would no doubt be suffering from severe drought conditions and a resulting smaller bread basket of their own. The great ice sheets that dominated the northern landmasses during past glacial periods would take a long time to reappear, but ice age conditions could persist globally in just decades.

New designs for human habitats, from small family dwellings to structures for large spacious communities, will be required to provide shelter during times of inclement weather. In time, as ice accumulates in the polar regions, sea levels will drop, exposing large new areas of fertile soils along all coastal margins. A house 150 m (nearly 500 ft) above sea level today, during glacial times, may be closer to 250 m (more than 800 ft) above the seas. Many coastal areas where today you can stand at the beach and enjoy the sights of the sea, during the last glacial period, would have been too far inland to see the sea. Initially, though, the amount of arable land will be much reduced compared to today, and the soils will be drier. New farming methods and improved freshwater supplies will be needed. The Sun, wind, and atom are all available to sustain us, but new and better ways to use them will be needed. The expected coming of fusion as a clean, safe, and inexhaustible energy source is undoubtedly one of the bright spots in today's world.

All of these things will be necessary to ensure that the basic needs of the world's still-growing population are met and to support its vast and rich diversity. This does not mean we must begin digging shelters or run out and buy extra blankets this afternoon. Nevertheless, the record of the past is our best indicator of where our climate is taking us today. The Earth's climate is inherently cyclical, and we can expect that it will continue to follow a natural and predictable path of change that it has followed many times before. Glacial conditions will eventually return, whether in centuries, mere decades, or just years from now.

If we think of a clock, as the image on this book's cover portrays, as representing the history, and the more ancient prehistory, of humanity on Earth, and the eleventh hour as representing our modern era, then we may today be counting down the final minutes before the climate turns cold again. It may not occur for a long time, and then again, it may begin this winter with the onset of colder temperatures and more severe changes in northern latitudes that will progressively build toward a permanent ice age climate- no one can say for sure.

The time will come when severe winter conditions will be the norm nearly everywhere in northern latitudes. Peter Wadhams, professor of ocean physics at Cambridge University, describing the severely weakened condition of the convection region in the Greenland Sea in a London Times editorial from a few years ago, said there were only a couple of weakly circulating columns of sinking water, or areas of convection, that did not even penetrate to the deep layer. This, compared to the nine to twelve giant columns of mixing water that would have once been found forming under a vast ice shelf- the Odden Ice Shelf- which no longer forms on the Greenland Sea due to the warmer winter conditions there. As stated previously, an interglacial climate cannot be maintained without deep-water convection in this region (Leake, 2005). The same is said about a change to cooling in the northeast Atlantic- that a decadal shift in this ocean process may be enough to initiate an ice age (Bischof, 2003). The acceleration of the Hydrological Cycle, which is the cause of many of our climate troubles, is more an aspect of glacial climate than an interglacial one.

What a contradiction the climate is these days! When this winter scenario comes about, life on Earth will have to adapt to the new conditions. The Dryas flower will likely become a common sight again as the Arctic tundra environment expands southward into the middle latitudes.

Initially, this may simply involve the use of good judgment, being prepared for inclement weather in all seasons, especially in winter, and taking extra precautions- when traveling, for instance. Naturally, the farther north one lives, the more important this will be. It means developing the same respect for winter weather that people living in the far north have always known. Governments should begin planning and preparing at community, state, provincial, and national levels. This event could very well occur in our lifetimes- or in the lives of our children or their children- and it is simply too important to ignore. The responsibility will principally fall upon governments as best able to implement necessary preparations. However, it will ultimately be up to everyone to help ensure that when the present Holocene warm period ends, no segment of the world's population is left out. We must all work together to ensure that when the music of the Holocene finally stops, no one is left without a chair.

Considerable knowledge has been gained about climate, especially over the last several decades, through the work of climatologists, oceanographers, geophysicists, and many other scientists in many related fields and many countries- it is their work on which this analysis is based. Even though science has advanced significantly in this field, we realize there is still much to learn. The ancient mariner setting out to sail across a sea could point the ship in a general direction, but the weather and ocean currents were changeable and not well understood. So, even with the aid of navigational instruments, the mariner could never foresee precisely the point of landfall. The voyage would always include some final course corrections toward the end. We are in a similar situation today with climate change. Important aspects of the climate system are identified. With the help of modern instruments like satellite technology, we have achieved at least a basic understanding of the individual components, how they interact with one another, and how they have changed in the past, thus making it possible to project a general direction that the climate is taking us. Like the ancient mariner, we know there will be some unknowns- some variables- along the way, and course corrections will be required from time to time. Like the ancient mariner, we can set a general course, even though we will not know, at the outset, our precise destination. We must not use our incomplete understanding as an excuse to delay preparations- we cannot afford to put this one off. It is unnecessary to have every piece of a puzzle to see the picture. The explorers of the past overcame their fears of the unknown, and we can, too. As though we have a choice in the matter, the climate will change when it is ready, regardless of what we do- we are going along, ready or not.

At the beginning of this interglacial period, after countless generations of ice age conditions, and as the climate warmed to temperatures like today, much of the warming happened well within a single human lifetime. Archaeological evidence shows that people quickly moved into newly greening areas at higher altitudes and latitudes as the ice receded. The Holocene has endured for some 11,500 years, which is a very long time, but when considering the long history of humanity on this planet, it is not really so long. Humankind has advanced during this era, from simple stone and metal tools to the space age. Comparatively speaking, we have come a long way very fast, due in no small part to the relatively comfortable climate conditions that have prevailed. Our present understanding of climate, how it has changed in the past, the changes occurring now, and the place we find ourselves in the glacial cycle presents us with a certain responsibility. When conditions return to a glacial climate, essential aspects of that change will likely happen relatively quickly. If we do nothing with this knowledge, the future may be mere survival. An unprepared world will be challenged to meet even the most basic societal needs. The recorded highlights of human history seem replete with war and suffering, yet there is so much to preserve. Much has been achieved; in the arts and the sciences, for example- and people are as spiritual today as they ever were. Ralph Waldo Emerson once wrote that success can be measured in knowing that "even one other life has breathed because you lived-" well, we can give life to many. We have it within us to build a great future for humanity on the Earth and beyond, regardless of how the story of climate eventually unfolds.

We, as good stewards of our planet

We must all become good stewards of our planet, for ourselves, our families, our neighbors, and fellow citizens- and for all the generations of people yet to be born. Future generations will have to adjust to living in a world where climate conditions will be less favorable, and in some regions of the world, much less favorable than those at present. Communities as they exist now in many places will not be sustainable. New emphasis is needed now in education to ensure that the physical and social sciences, engineering, and agriculture have the bright minds needed to help create future sustainable and livable communities. New curricula should be developed in universities with this single goal: to get these people working where they can make a difference.

This book, of course, is not the last word on climate change; a final solution must include a consensus of many ideas and opinions- but let us get started. May it be said by future generations that this was a time when a new course was set, that this was a turning point when a great fire was lit that inspired the hearts of humanity- and that the glow from that fire grew to light the way toward a brighter world. So, let us begin; there is no greater challenge before us and no greater gift we can give.

Amid today's warming, a cold-weather clock ticks away in the northern seas. The climate in the North Atlantic region has taken a couple of important steps toward colder climate conditions. The great Atlantic Current that has fueled this time of warming is declining in influence, and a cold Arctic Blob now dominates the North Atlantic in its place. The strengthening occurring in the Earth's Hydrological Cycle, with the drying of the lands, the accumulation of moisture in the atmosphere, and the associated increase in atmospheric processes, is significant as a global indicator of change. These are considered long-term changes, and we should recognize them as an early signal to begin preparations for the return of the climate to glacial conditions, for the return of winter- for *The Coming of the Dryas.*

Early 70's-
Time of the 'Great Salinity Anomaly'

1970-
Intense deep water convection in the Greenland Sea, while convection in the Labrador Sea is shallow

Early 50's-
A cooling trend begins in the surface layer of the northeast Atlantic

Early 60's-
Decreasing temperatures in northern Scandinavia and Russia

1972-
Begins large increase of icebergs in the northwest Atlantic

1953-
Increased inflow of Arctic fresh water discovered in the Subpolar Seas

1963-
Norwegian maritime glaciers begin new advancement

1973-
Greenland Sea begins to warm- convection process begins to weaken

1950   1960   1970

1954/55-
A cooling begins in the lower Stratosphere

Late 60's-
Great increase in the southward flow of Arctic fresh water- extending into the North Atlantic and beyond

Labrador Sea begins to freshen

Mid-to-late 70's-
Begins an almost linear freshening of the Nordic Seas, and a doubling of the overflow of south-bound water across the Greenland-Scotland Ridge

Timeline of change/NAO numerical index

Early-to-Mid 90's-
A large surge of fresh water and ice into the Nordic Seas and the Subpolar North Atlantic

Strong deep-water convection in the Labrador Sea

1993-
Onset of intense Stratospheric cooling

1994-
Virtual cessation of deep-water convection in the Greenland Sea

Weakening of convection in the Labrador Sea

Early 2000s-
Warming intensifies in the surface layer of the subpolar region- ocean circulation continues to weaken

Rapid retreat of Norwegian glaciers

Only intermittent deep-water convection in the Greenland and Labrador Seas

1980-
Dramatic & sustained weakening of convection begins in Greenland Sea

1980 — **1990** — **2000**

1986-
Start of significant warming in Norwegian Sea deep layer

1987-
Acceleration of Arctic melting

Significant advance of Norwegian glaciers

1996-
Most intense winter NAO negative phase on record

Begins weakening and contraction of the North Atlantic circulation system

Begins sharp decline of convection in the Labrador Sea

Strong winter cooling and freshening in the subpolar surface layer gives way to new trend toward warmer and saltier conditions

New dominance of cold water sea plankton

1998-
Significant weakening discovered in the Deep Atlantic Current

2004/05-
Dramatic warming and slowing evident in the Norwegian Current

2009-
Begins significant weakening of the North Atlantic Current

Increasing Arctic water in Western Subtropical Gyre

2010-
New Record lows in
NAO- unprecedented
negative trend in all
significant indexes

Dramatic 30% decline
in the North Atlantic
Current.

Increased recirculation
of Atlantic water at
depth through the
Thermocline

2011-
Atlantic Current
recovers strength-
begins steady decline

Early 2020s-
Cooling trend continues
in North Atlantic and
Nordic Seas

Sea-ice volume holding
near 2010 levels

?
The next
climate shift

2010   2020   2030

2015-
Significant cooling &
freshening of North
Atlantic. Arctic water
('Atlantic Cold Blob')
achieves dominance

Weakened Atlantic
Current relegated to
narrowed pathway
along Blob's eastern
margin.

2017-
50% reduction of
coastal glacial melting
in Greenland, Iceland
and Scandinavia

Jakobshavn, region's
largest glacier, begins
to grow

NAO annual/seasonal (numerical) indexes
(second column is 5 year running average)

	Annual		Summer		Winter	
1990	1.23	0.30	1.04	-0.11	2.37	0.99
1991	0.34	0.26	0.25	0.13	0.20	1.17
1992	1.11	0.58	1.06	0.52	1.68	1.44
1993	0.12	0.67	0.26	0.56	1.42	1.70
1994	0.51	0.66	-0.38	0.44	1.80	1.49
1995	-0.61	0.29	-1.80	-0.12	2.44	1.50
1996	*-1.01	0.02	0.12	-0.14	*-2.32	1.00
1997	-0.18	-0.23	-0.44	-0.44	0.18	0.70
1998	0.25	-0.20	-0.77	-0.65	0.80	0.58
1999	0.04	0.30	-1.16	-0.81	0.98	0.41
2000	0.03	-0.17	-0.60	-0.39	1.85	0.30
2001	-0.45	-0.06	-1.16	-0.65	-0.50	0.66
2002	-0.04	-0.13	-1.00	-0.94	0.79	0.78
2003	-0.30	-0.14	-0.35	-0.85	0.40	0.70
2004	0.00	-0.15	0.27	-0.57	-0.20	0.47
2005	-0.25	-0.21	0.09	-0.43	-0.11	0.07
2006	-0.20	-0.16	-0.76	-0.35	-0.82	0.01
2007	-0.38	-0.23	*-2.24	-0.60	1.83	0.22
2008	-0.73	-0.31	-1.43	-0.81	1.37	0.41
2009	-0.41	-0.39	-0.88	-1.04	-0.31	0.39
2010	*-2.18	-0.78	-1.68	-1.39	*-2.54	-0.09
2011	0.46	-0.65	-1.10	-1.47	-0.84	-0.10
2012	-0.63	-0.70	-1.44	-1.31	2.07	-0.05
2013	0.59	-0.43	1.38	-0.74	-0.58	-0.44
2014	0.10	-0.33	-1.08	-0.78	2.05	0.03
2015	1.16	0.34	-0.51	-0.55	2.04	0.95
2016	0.39	0.32	1.38	-0.05	1.83	1.48
2017	0.57	0.56	0.23	0.28	1.17	1.30
2018	0.72	0.59	1.03	0.21	0.38	1.49
2019	-0.03	0.56	-1.46	0.13	1.46	1.38
2020	0.33	0.40	-0.90	0.06	-0.28	0.91
2021	-0.67	0.18	-0.96	-0.41	-0.08	0.53
2022	-0.06	-0.07	-1.29	-0.72	1.21	0.54

* Years of particular interest

The measure of the North Atlantic Oscillation (NAO) is expressed as either a positive or negative numerical value. A positive phase brings in generally warmer and milder atmospheric conditions; heat from the northward-moving Atlantic Current is picked up by eastward-moving winds, which carry that warmth across Europe and into western Asia. However, when the NAO changes to a negative phase, those winds shift farther south, and weather conditions in the North Atlantic and surrounding lands become colder and stormier.

There is a continuous record of NAO measurements beginning in 1825. Here, we examine the years 1990 to 2023, when some larger changes began to occur, with a focus on those numbers that exceed a whole numerical value- in other words, more than 1.00 or less than -1.00. Looking at the record from an annual and seasonal perspective, there are some important clues to the direction in which things are headed.

On the Annual Index, there are only three years from 1850 to 1990 where a whole numerical value is exceeded, and all three are positive- which means generally mild weather conditions. On the Summer Index- including both positive and negative seasons- the average is about 1.5 per decade in that same period. In the two decades following 1990, that number more than doubled for the same Summer Index. The year 2007, however, marked the beginning of a truly unprecedented era, for those numbers have, since then, doubled yet again.

Winters, of course, are always the most extreme with regard to weather conditions, and for the NAO, there is no exception. The number of winter seasons for those years prior to 1990, where that whole number value was exceeded, averaged 3.6 per decade. There has been an increase in the following decades, although not so remarkable as in the Annual and Summer Indexes. What is remarkable, however, is the direction and intensity of the winters during two of those years. The Annual Index exceeded a negative whole numerical value (-1.01) for the first time on record in 1996. The winter that year was not the first to fall into negative numbers on the index, but it set a new record low at -2.32, which was exceeded fourteen years later in 2010 at -2.54.

What stands out on these indexes after 1990, as compared to the preceding 140 years since 1850, aside from those two extreme winters, are the wild oscillations in climate conditions that have begun in the region. These are particularly evident in the Winter Index, which is likely the first place we look for clues to expected increases. Similar oscillations preceded the onset of the Younger Dryas cold period in the Early Holocene. A persistent negative trend evident in the Summer Index since 2010 is of particular concern. As we have seen, the climate record clearly reveals deteriorating summer conditions as one of the markers that preceded past transitions to a glacial climate.

TERMS & DEFINITIONS

8,200 Yr. Event: A 200-year long temperature decline that occurred during the early Holocene.

Arctic/Atlantic Water Balance: Describes the relative abundance of Arctic and Atlantic derived water in the North Atlantic region.

Arctic Oscillation: Primary indicator for weather and climate conditions in the Northern Hemisphere.

Atlantic Cold Blob: Very large mass of cold fresh Arctic derived water beginning after 2010, centered in the Eastern Subpolar Gyre, and now dominant in the North Atlantic.

Atlantic Storm Tract: Atmospheric frontal system extending across the North Atlantic to Europe, generally defined by the boundaries of the North Atlantic Current.

Axial Inclination: Refers to a long period- 41,000 year- cycle, or change, in the tilt of the Earth's axis- a primary determinant in global climate.

Axial Inclination Maximum: The point in the Axial Inclination Cycle where the Earth's axis is at its greatest angle toward the Sun in relation to the plane of its orbit. This point was reached 7,000 years ago, at the height of Early Holocene warming.

Biomass Burning: Large increase in the occurrence of wildfires, or woodland burning, apparent at the beginning of the Younger Dryas Cold Event- and which is also apparent today in many parts of the world.

Bolling-Allerod: Relative warm period following the end of the Pleistocene glaciation, and preceding the Younger Dryas.

Climate Record: The palaeontological, or geological record, of past climate conditions.

Climate Transition: The changing from one climate state to an alternate one. Important aspects of transitions will complete in just decades or less, and are global in their effect.

Convection & Formation: In each of the Subpolar Seas surface water sinks- called convection- to at least intermediate depths and begins the formation of underwater currents that return south, completing the Atlantic link in ocean circulation.

Daytime (diurnal) Temperature Range (DTR): The range within which daytime temperatures tend to occur. Part of what is called 'Global warming' actually involves an increase of average temperature within this range.

Deep Water Current (the Deep Current): Lower North Atlantic Deep Water (LNADW), the Atlantic link in global ocean circulation.

Deep Western Boundary Current (DWBC): Consisting of deep southbound currents, (LNADW) and (UNADW) from the Greenland and Labrador Seas.

Eemian & Penultimate interglacial periods: The two most recent periods of interglacial warming.

Glacial Cycle: Oscillation of the climate between interglacial periods of relative warmth and colder glacial conditions- the dominant trend in the Earth's climate for some 1.5 million years.

Holocene Interglacial Period: The time of interglacial warmth in which we now live. It began with conditions warmer than today- the 'Early Holocene-' which culminated in a time of maximum warmth, called the 'Mid-Holocene Optimum,' followed by a 'Late Holocene' period of cooler and drier climate.

Hydrological Cycle: Also called the water cycle, it involves the cycling, of moisture (water) through the climate system. On a global scale it involves the transfer of warm tropical air and moisture to the cold polar regions.

Iceland Basin: Relatively shallow ocean region in the eastern Subpolar North Atlantic- not the same as the Iceland Sea. The Atlantic Current mixes here with water of Norwegian Sea origin.

Insolation: Warmth created by the focus of direct sunlight on the Earth's surface. It follows with north-to-south changes in the precessional cycle.

Latitude: Indicates a north-to-south geographical location on the Earth's surface. Beginning with 0^0 at the equator and progressing to 90^0 at the poles; the term 'lower latitude' indicates the tropical and subtropical regions, high latitudes are the polar regions, and the middle latitudes those areas between.

Little Ice Age : Coldest era of the Holocene; reaching its height some 300 years ago, only just ending with the current period of warming.

Medieval Warm Period: The time of warmth preceding the Little Ice Age; warmer than today in many areas, it endured for nearly 700 years in Europe.

Modes of Ocean Circulation: Three basic modes of ocean circulation in the North Atlantic region are identified as directly associated with prevailing glacial or interglacial climate states.

Nordic Seas: With Norway to the east and Greenland to the west, and the Arctic Ocean to the north, and defined by the Greenland-Scotland Ridge to the south, this area encompasses the Greenland, Iceland, and Norwegian Seas.

North Atlantic Current: A great northward flowing current of tropical origin; its strength is essential to the continuation of our warm interglacial climate- farther south it is called the Gulf Stream.

North Atlantic Drift Current: A wide swath of slowly moving surface water in the far northeast Atlantic. Its warming and weakening in recent decades is a major point of concern.

North Atlantic Oscillation (NAO): A primary component of the Arctic Oscillation, it is the most important indicator for climate conditions in the North Atlantic region.

North Atlantic Region/ Northern Seas: In this study, indicates the area encompassing all of the North Atlantic and Nordic Seas.

Ocean Circulation: Great ocean currents circulate at depth and at the surface thorough all of the oceans, connected by sometimes giant convective regions of sinking or risings columns of mixing water- the Atlantic link in this system being the largest of these.

Orbital Eccentricity: A major climate parameter; the shape of the Earth's orbit changes over a very long period of time from nearly circular to a more elliptical one.

Perihelion: Point in the Earth's orbit where it is closest to the Sun.

Pleistocene Ice Age: The one thousand centuries of cold ice-age conditions that prevailed prior to the advent of the present Holocene Interglacial Period.

Polar Front: Atmospheric and oceanic frontal system, or boundary, that, in the Nordic Seas, separates Arctic derived water from that of Atlantic origin.

Precession: Perhaps the most important of orbital parameters, it involves a rotation of the earth's axis, over an approximately 23,000-year period, opposite the direction of its revolution and rotation. An apparent likewise motion of the Sun and seasons along the ecliptic is referred to as the 'Precession of the Equinox.'

Precessional Maximum: Refers to the point in the Precessional Cycle where the Earth is at the same time at its closest approach and its axis is at its most direct (precessional) angle toward the Sun.

Subpolar Gyre: Consisting of an eastern and western region of circulating ocean water within the Subpolar North Atlantic, its slowing in recent decades is an important aspect of change.

Subpolar North Atlantic: Stretching from Ireland to Canada, this is the area of the far North Atlantic.

Subpolar Seas: Includes all of the Nordic Seas- the Greenland, Iceland and Norwegian Seas- plus the Labrador and Irminger Seas in the western subpolar region.

Subtropical Re-circulation: Refers to a large and growing quantity of Atlantic water that has been observed in recent decades to be breaking away from the North Atlantic Current and recirculating back into the subtropical Atlantic.

Younger Dryas: The most recent of major cold events, it was initiated by a sudden influx of fresh water from melting continental ice sheets on North America- returning the climate to glacial conditions for a thousand years.

REFERENCES

Adams, J., Maslin, M., Thomas, E., 1999: Sudden Climate Transitions During the Quaternary, Progress in Physical Geography, 23(1), 1-36.

Alkire, Matthew B., [1]Jacobson, Andrew, [2]& Macdonald, Robie W., [3]20 May 2019: Assessing the Contributions of Atmospheric/ Meteoric Water and Sea Ice Meltwater and Their Influences on Geochemical Properties in Estuaries of the Canadian Arctic Archipelago. [1]Applied Physics Laboratory, University of Washington, Seattle, WA, USA, [2]Department of Earth & Planetary Sciences, Northwestern University, Evanston, IL, USA, [3]Department of Fisheries & Oceans, Institute of Ocean Sciences, Sidney, BC, Canada, Estuaries and Coasts (2019) 42:1226–1248

Alley, R., 2004: GISP2 Ice Core Temperature and Accumulation Data. IGBP PAGES/ World Data Center for Palaeoclimatology Data Contribution Series #2004-013. NOAA/NGDC Palaeoclimatology Program, Boulder CO, USA.

Anderson, C., Koc, N., Moros, M., 2004: A Highly Unstable Holocene Climate in the Subpolar North Atlantic: Evidence From Diatoms, Quaternary Science Reviews, 23, 2155-2166.

Andreassen, L., Elvehøy, H., Kjøllmoen, B., 10 May 2005: Major Changes in Norway's Glaciers, Center for International Climate and Environmental Research (CICERO).

Astronomical Theory of Climate Change, 6 April 2009: US National Oceanic and Atmospheric Administration (NOAA), Satellite and Information Service, National Climate Data Center, Paleoclimatology.

Bischof, B., Mariano, A., Ryan, E., 2003: The North Atlantic Drift Current, Surface Currents in the Atlantic Ocean, The Cooperative Institute for Marine and Atmospheric Studies (CIMAS).

Box, J., Yang, L., Bromwich, D., 15 July 2009: Greenland Ice Sheet Surface Air Temperature Variability: 1840-2007, Journal of Climate, 22, 4029-4049.

Broecker, W., May 1997: Will Our Ride Into the Greenhouse Future Be a Smooth One? GSA Today, 7(5).

Broecker, W., 27 November 1997: Columbia Scientist Warns That Global Warming Could Trigger Collapse of Ocean Currents: Halt to Gulf Stream and Other Currents Could Freeze Europe; Dublin Would Share Spitsbergen's Icy Climate, Columbia University News.

Broecker, W., 1999: What if the Conveyor Were to Shut Down? Reflections on a Possible Outcome of the Great Global Experiment, GSA Today (Geological Society of America) 9(1),1-7.

Bryden, H., Longworth, H., Cunningham, S., 2005: Slowing of the Atlantic Meridional Overturning Circulation at 25° N. Nature, 438, 655-657.

Bryden, Harry L., Johns, William E., King, Brian A., 1 March 2020: Reduction in Ocean Heat Transport at 26°N since 2008 Cools the Eastern Subpolar Gyre of the North Atlantic Ocean. J. Climate, **33**, 1677-1689.

Buis, A., Chambers, D., Hilburn, K., 4 October 2010: Nasa Study Sees Earth's Water Cycle Pulse Quickening, US National Atmospheric and Oceanic Administration (NASA), Jet Propulsion Laboratory.

Calvin, W., 1991: Greenhouse and Hothouse, Whole Earth Review, 73, 106-111.

Calvin, W., January 1998: The Great Climate Flip-flop, The Atlantic Monthly, 281(1), 47-64.

Castelvecchi, Davide, 26 February 2020: Mysterious faded star Betelgeuse has started to brighten again. Nature, Space & Physics.

Clark, P., Pisias, N., Stocker, T., 21 February 2002: The Role of the Thermohaline Circulation in Abrupt Climate Change, Nature, 415, 863-869.

Claussen, M., Kubatzki, C., Brovkin, 15 July 1999: Simulation of an Abrupt Change in Saharan Vegetation in the Mid-Holocene, Geophysical Research Letters, 26(14), 2037-2040.

Cottet-Puinel, M., Weaver, A., Hillaire-Marcel, C., 2004: Variation of Labrador Sea Water Formation Over the Last Glacial Cycle in a Climate Model of Intermediate Complexity, Quaternary Science Reviews, 23, 449-465.

Cronin, T., 1999: Paleo-atmospheres- The Ice-core Record of Climate Change, Principles of Paleoclimatology, Perspectives in Paleobiology and Earth History, Columbia University Press.

Curry, R., 16 June 2005: How Much Excess Fresh Water Was Added to the North Atlantic in Recent Decades? Continued Freshening of the North Atlantic Could Slow the Conveyor in the 21st Century, News Release, Woods Hole Oceanographic Institution.

deMenocal, P., Ortiz, J., Guilderson, T., 23 June 2000: Coherent High- and Low-Latitude Climate Variability During the Holocene Warm Period, Science, 288, 2198-2202.

Dickson, B., Yashayaev, I., Meincke, J., 25 April 2002: Rapid Freshening of the Deep North Atlantic Ocean Over the Past Four Decades, Nature, Letters to Nature, 416, 832-836.

Drange, H., Dokken, T., Furevik, T., 2005: The Nordic Seas: An overview, in The Nordic Seas: An Integrated Perspective (Drange, Dokken, Furevik, Eds.), AGU Monograph 158, American Geophysical Union, Washington DC.

Drought's Growing Reach: NCAR Study Points to Global Warming as Key Factor, 10 January 2005: University Corporation for Atmospheric Research (NCAR), Press Release.

Eiriksson, J., Bartels-Jonsdottir, H., Cage, A., 2006: Variability of the North Atlantic Current During the Last 2000 Years Based on Shelf Bottom Water and Sea Surface Temperatures Along an Open Ocean -Shallow Marine Transect in Western Europe, The Holocene, 16(7), 1017-1029.

Elias, S., Schreve, D., 2007: Late Pleistocene Megafaunal Extinctions, Vertebrate Records, 3202-3217.

Esselborn, S., Miller, L., Cheney, B., 1999: Interannual Changes in North Atlantic Sea Level and Surface Circulation as Measured by Satellite Altimetry. US Department of Commerce. National Oceanic and Atmospheric Administration (NOAA)

Gagosian, R., 27 January 2003: Abrupt Climate Change, Should We Be Worried? Woods Hole Oceanographic Institution, World Economic Forum.

Greenland Glacier Gives Birth to Giant Iceberg, 9 August 2010: European Space Agency (ESA), Observing the Earth.

Greiner, Benkiran, Pergaud, 2006: The Mercator 1992-2002 PSY1V2 Reanalysis for Tropical and North Atlantic. Mercator-Ocean, Toulouse, France.

Häkkinen, S., and Rhines, P., 2004: Decline of Subpolar North Atlantic Circulation During the 1990s, Science, 304, 555-559.

Hanssen-Bauer, I., Førland, E., 14 August 1998: Long-term Trends in Precipitation and Temperature in the Norwegian Arctic: Can They Be Explained Changes in Atmospheric Circulation Patterns, Climate Research, 10, 143-153.

Hansen, B., Turrell, W., Østerhus, S., 21 June 2001: Decreasing Overflow From the Nordic Seas Into the Atlantic Ocean Through the Faroe Bank Channel Since 1950, Nature, 411, 927-930.

Hátún, H., Sando, A., Drange, H., 2005: Influence of the Atlantic Subpolar Gyre on the Thermohaline Circulation, Science, 309, 1841-1844.

Holliday, N., Bersch, M., Berx, B., 2020: Ocean circulation causes the largest freshening event for 120 years in eastern subpolar North Atlantic. nature communications, 11:585.

Hotz, R., 7 July 2006: "Wildfire Increase Linked to Climate," Times Staff Writer, Los Angeles Times.

How Fast Did Climate Change During the Glacial Period? 2001: Working Group I: The Scientific Basis, Intergovernmental Panel On Climate Change.

Hurrell, J., 1995: Decadal trends in the North Atlantic Oscillation and Relationships to Regional Temperature and Precipitation, Science, 269, 676-679.

International Ice Patrol South of 48 N Latitude Data Set Since 1900, 2009: US Coast Guard, International Ice Patrol Archived Data.

Jones, P., Jónsson, T. and Wheeler, D., 1997: Extension to the North Atlantic Oscillation Using Early Instrumental Pressure Observations from Gibraltar and South-West Iceland. Int. J. Climatol. 17, 1433-1450.

Joyce, T., 18 April 2002: "The Heat Before the Cold," Woods Hole Oceanographic Institution, The New York Times.

Kakad, Amar, Kakad, Bharati, 1 March 2022: An audit of geomagnetic field in polar and south atlantic anomaly regions over two centuries. Advances in Space Research, Volume 69, Issue 5, pages 2142-2157

Karstensen, J., Schlosser, P., Wallace, D., 27 July 2005: Water Mass Transformation in the Greenland Sea During the 1990's, Journal of Geophysical Research, 110, C07022, doi:10.1029/2004JC002510.

Kerr, R., 16 April 2004: A Slowing Cog In the North Atlantic Ocean's Climate Machine, Science, Global Change, News of the Week, 304.

Khatiwala, S., Schlosser, P., and Visbeck, M., February 2002: Rates and Mechanisms of Water Mass Transformation in the Labrador Sea as Inferred from Tracer Observations, Journal of Physical Oceanography, 32, 666- 686.

Koc, N., Jansen, E., Haflidason, H., 1993: Paleoceanographic Reconstructions of Surface Ocean Conditions in the Greenland, Iceland and Norwegian Seas Through the Last 14ka Based on Diatoms, Quaternary Science Reviews, 12(2), 115-140.

Koc, N., 2-6 June 2000: Reconstructing the Sea Ice Limit in the Nordic Seas Through the Last 14ka Using Diatom Sea Ice Assemblages, Sea Ice in the Climate System, The Record of the North Atlantic Arctic, Circumpolar Arctic Paleo-Environments

Leake, J., 8 May 2005: "Britain Faces Big Chill as Ocean Current Slows." Editorial, The Sunday Times (London).

Lozier, M. S., Li, F., Bacon, S., 1 February 2019: A sea change in our view of overturning in the subpolar North Atlantic. Science 363, 516–521.

Marsh, R., 2000: Recent Variability of the North Atlantic Thermohaline Circulation Inferred From Surface Heat and Freshwater Fluxes, Journal of Climate, 13(18), 3239-3260.

McBean, G., Alekseev, G., Deliang, C., 9-12 November 2004: Arctic Climate: Past and Present, Arctic Climate Impact Assessment (ACIA), AMAP Report 2004: The ACIA International Scientific Symposium on Climate Change in the Arctic, Extended Abstracts, Reykjavik, Iceland 2, 47.

McCarthy, G., and Coauthors, 2012: Observed interannual variability of the Atlantic meridional overturning circulation at 26.58N. Geophys. Res. Lett., 39, L19609

Milankovitch Variations.png, 11 February 2006: Wikipedia The Free Encyclopedia. <http://www.en.wikipedia.org/wiki/File: Milankovitch Variations.png> (5 March 2012).

Milutin Milankovitch (1879-1958), Orbital Variations, 11 March 2011: US National Aeronautics and Space Administration (NASA), Earth Observatory. <http://www.earthobservatory.nasa.gov/ Features/ Milankovitch/milankovitch_2.php> (5 March 2012).

Mrugala, R., 2012: NOAA Paleoclimatology Program (graphic images). <http://www.ncdc.noaa.gov/paleo/sciencepub/ figure3.htm> (19 March 2012).

New Study Reports Large-Scale Salinity Changes in the Oceans: Saltier Tropical Oceans and Fresher Ocean Waters Near the Poles Further Signs of Global Climate Change's Impacts, 17 December 2003: National Science Foundation, Office of Legislative and Public Affairs, Press Release. <http://nsf.gov/od/lpa/news/03/ pr03145.htm> (18 April 2012).

Noël, Brice, [1]Aðalgeirsdóttir, Guðfinna, [2]Pálsson, Finnur,[2] 14 JAN 2022: North Atlantic Cooling is Slowing Down Mass Loss of Icelandic

Glaciers. [1]Institute for Marine and Atmospheric Research Utrecht, Utrecht University, Utrecht, The Netherlands, [2]Institute of Earth Sciences, University of Iceland, Reykjavìk, Iceland, Geophysical Research Letters, 10.1029/2021GL095697.

North Atlantic Oscillation (NAO), 18 July 2011: UK Climate Research Unit, University of East Anglia.

Olafsson, J., Ostermann, D., Curry, W., 2000: Time Series Particle Fluxes from the Iceland Sea: 1986 to Present, Woods Hole Oceanographic Institution.

Orvik, K., Skagseth, Ø., 5 December 2005: The Norwegian Atlantic Current Has Become Warmer and Weaker Over the Last 10 Years, Center For International Climate and Environmental Research (CICERO).

Osborn, T., 2000: North Atlantic Oscillation, UK Climate Research Unit, University of East Anglia, Information Sheets. <http://www.cru.uea. ac.uk/cru/info/nao/> (10 February 2012).

Osborn, T., Briffa, K., Schweingruber, F., June 2004: Annually Resolved Patterns of Summer Temperature Over the Northern Hemisphere Since AD 1400 From a Tree-Ring Density Network, UK Climate Research Unit, University of East Anglia, Swiss Federal Institute of Forest, Snow and Landscape Research, Submitted to Global and Planetary Change.

Østerhus, S., Gammelsrød, T., Hogstad, R., 14 May 1997: Ocean Weather Ship Station M (66°N, 2°E): The Longest Existing Homogeneous Time Series From the Deep Ocean.

Østerhus, S., Gammelsrød T., November 1999: The Abyss of the Nordic Seas is Warming, Journal of Climate 12, 3297-3304.

Ostermann, D., Curry, W., Honjo, S., 2000: Time Series Foraminiferal Fluxes from the Iceland Sea: 1986 to Present, Woods Hole Oceanographic Institution.

Petit, J., Jouzel, J., Raynaud, D., 1999: Climate and Atmospheric History of the Past 420,000 years from the Vostok Ice Core, Antarctica, Nature, 399, 429-436.

Rimbu, N., Lohmann, G., Lorenz, S., 22 June 2004: Holocene Climate Variability as Derived from Alkenone Sea Surface Temperature and Coupled Ocean-Atmosphere Model Experiments, Climate Dynamics, 23, 215-237

Sample, I., 30 November 2005: "Alarm Over Dramatic Weakening of Gulf Stream," The Guardian.

Satellites Record Weakening North Atlantic Current, 15 April 2004: US National Atmospheric and Space Administration (NASA), Goddard Space Flight Center.
<http://www.nasa.gov/centers/goddard/news/topstory/2004/0415gyre.html> (10 February 2012).

Schaeffer, M., Selten, F. Opsteegh, J., 2002: Intrinsic Limits to Predictability of Abrupt Regional Climate Change in IPCC SRES Scenarios, Geophysical Research Letters, 29(0).

Smeed, D. A., [1]Josey, S. A., [1]Beaulieu, C., [2]12 FEB 2018: The North Atlantic Ocean Is in a State of Reduced Overturning. [1]National Oceanography Centre, Southampton, UK, [2]Ocean and Earth Science, University of Southampton, Southampton, UK, Geophysical Research Letters,10.1002/2017GL076350.

Sokov, A., Falina, A., 2006: Interannual Variability of the Water-Mass Properties in the Subpolar North Atlantic at 60°N, 1997 -2005, American Geophysical Union, 87(36), Ocean Science Meeting Supplement.

Special Climate Summary- 96/1: Climate Conditions During the 1995/96 Northern Hemisphere Winter, March, 1996, US National Weather Service, National Centers for Environmental Prediction, Climate Prediction

Center, Climate Operations Branch & Analysis Branch. <http://www.cpc. ncep.noaa.gov/products/special summaries/96_1/> (8 February 2012).

Stein, M., 2005: North Atlantic Subpolar Gyre Warming- Impacts on Greenland Offshore Waters, J. Northw. Alt. Fish. Sci., 36, 1-12.

Taylor, G., March 2006: Antarctic Temperature and Sea Ice Trends Over the Last Century, An Assessment of Antarctic Climate Data.

Taylor, K., July-August 1999: Rapid Climate Change: New Evidence Shows That Earth's Climate Can Change Dramatically in Only a Decade. Could Greenhouse Gases Flip That Switch? American Scientist.

The Sunspot Cycle, 2012: US National Aeronautics and Space Administration NASA), Solar Physics, Marshall Space Flight Center.

Tropospheric and Lower Stratospheric Temperature Anomalies Based on Global Radiosonde Network Data, CDIAC/Trends Online, 2012: US National Aeronautics and Space Administration (NASA), Goddard Space Flight Center, Global Change Master Directory, Discover Earth Science Data and Services. <http://www.gcmd.nasa.gov/KeywordSearch/ Metadata.do? Portal....> (19 March 2012).

Tuomenvirta, H., Alexandersson, H., Drebs, A., 5 March 2000: Trends in Nordic and Arctic Temperature Extremes and Ranges, Journal of Climate, 13(5), 977-990.

Volkov, D., van Aken, H., 22 July 2005: Climate Related Change of Sea Level in the Extratropical North Atlantic and North Pacific in 1993-2003, Geophysical Research Letters, 32(10), 1029.

Vörösmarty, C., Hinzman, L., Peterson, B., 2001: 4. Unprecedented Change to Arctic Hydrological Systems, The Hydrologic Cycle and its Role in Arctic and Global Environmental Change: A Rationale and Strategy for Synthesis Study, Arctic Research Consortium of the U.S.

Witak, M., Wachnicka, A., Kuijpers, A., 2005: Holocene North Atlantic Surface Circulation and Climate Variability- Evidence from Diatom Records, The Holocene 15(1), 85-96.

Ye, Aihua, Zhu, Zhipeng, Zhang, Ruyi, 19 May, 2023: Influence of solar forcing on multidecadal variability in the Atlantic meridional overturning circulation (AMOC). Earth Sci., Sec. Interdisciplinary Climate Studies, Volume 11.

INDEX